WHAT IS LIFE?

"By formulating a new stability kind in nature, Addy Pross has uncovered the chemical roots of Darwinian theory, thereby opening a novel route connecting biology to chemistry and physics. That connection suggests that abiogenesis and biological evolution are one process, throwing exciting new light on the origin of life and offering a striking chemical explanation for life's unusual characteristics. This book is more than worth reading—it stirs the readers' mind and paves the way toward the birth of further outstanding ideas."

Ada Yonath, *joint winner of the Nobel Prize for Chemistry, 2009*

What is Life?
HOW CHEMISTRY BECOMES BIOLOGY

ADDY PROSS

OXFORD
UNIVERSITY PRESS

Great Clarendon Street, Oxford, OX2 6DP,
United Kingdom

Oxford University Press is a department of the University of Oxford.
It furthers the University's objective of excellence in research, scholarship,
and education by publishing worldwide. Oxford is a registered trade mark of
Oxford University Press in the UK and in certain other countries

© Addy Pross 2012

The moral rights of the author have been asserted

First Edition published in 2012

Impression: 2

All rights reserved. No part of this publication may be reproduced, stored in
a retrieval system, or transmitted, in any form or by any means, without the
prior permission in writing of Oxford University Press, or as expressly permitted
by law, by licence or under terms agreed with the appropriate reprographics
rights organization. Enquiries concerning reproduction outside the scope of the
above should be sent to the Rights Department, Oxford University Press, at the
address above

You must not circulate this work in any other form
and you must impose this same condition on any acquirer

British Library Cataloguing in Publication Data
Data available

Library of Congress Cataloging in Publication Data
Data available

ISBN 978–0–19–964101–7

Printed and bound by
CPI Group (UK) Ltd, Croydon, CR0 4YY

To Nella, Guy, and Tamar, for what my life is

CONTENTS

Prologue	viii
1. Living Things are so Very Strange	1
2. The Quest for a Theory of Life	32
3. Understanding 'Understanding'	43
4. Stability and Instability	58
5. The Knotty Origin of Life Problem	82
6. Biology's Crisis of Identity	111
7. Biology is Chemistry	122
8. What is Life?	160
References and Notes	193
Index	199

PROLOGUE

'I spent the afternoon musing on Life. If you come to think of it, what a queer thing Life is! So unlike anything else, don't you know, if you see what I mean.'

PG Wodehouse

The subject of this book addresses basic questions that have transfixed and tormented humankind for millennia, ever since we sought to better understand our place in the universe—the nature of living things and their relationship to the non-living. The importance of finding a definitive answer to these questions cannot be overstated—it would reveal to us not just who and what we are, but would impact on our understanding of the universe as a whole. Has the universe been fine-tuned to support life, as implied by proponents of the so-called anthropic principle? Or, to take a more Copernican view of man's place in the universe, 'is the human race just a chemical scum on a moderate-sized planet', as argued by Stephen Hawking, the noted physicist? A wider conceptual gulf would be hard to conceive.

Some 65 years ago another renowned physicist, Erwin Schrödinger, wrote a book whose catchy title *What is Life?* directly addressed the issue. In the opening lines of that book Schrödinger wrote:

> How can the events in space and time which take place within the spatial boundary of a living organism be accounted for by physics and chemistry? The preliminary answer...can be summarized as follows: The obvious inability of present-day physics and chemistry

to account for such events is no reason at all for doubting that they can be accounted for by those sciences.

Sixty-five years have passed but despite the enormous advances in molecular biology in those years, illuminated by a long list of Nobel prizes, we continue to struggle with Schrödinger's simple and direct question. And a struggle it is. Carl Woese, one of the leading biologists of the twentieth century, has recently gone as far as to claim that the state of present-day biology is reminiscent of that of physics at the turn of the twentieth century, before Albert Einstein, Niels Bohr, Erwin Schrödinger, and the other great twentieth-century physicists totally revolutionized the subject; that the time for biology's revolution has finally come. Strong sentiments indeed! What is no less remarkable is that modern biology appears to be happily meandering along its current mechanistic path with most of its practitioners indifferent, if not oblivious, to the shrill cry for reassessment.

Yes, it is true that in this modern era we know unequivocally that there is no *élan vital*, that living things are made up of the same 'dead' molecules as non-living ones, but somehow the manner in which those molecules interact in a holistic ensemble results in something very special—us, and every other living thing on this planet. So, paradoxically, despite the profound advances in molecular biology over the past half-century, we still do not understand what life is, how it relates to the inanimate world, and how it emerged. True, over the past half-century considerable effort has been directed into attempts to resolve these fundamental issues, but the gates to the Promised Land seem as distant as ever. Like a mirage in the desert, just as the palm trees signalling the oasis seemingly materialize, shimmering on the horizon, they fade away yet again, leaving our thirst to understand unquenched, our drive to comprehend unsatisfied.

So what is the basis of this deeply troubling and persistent dilemma? To clarify in simplest terms where the problem lies, consider the following hypothetical tale: you are walking through a field and you suddenly come across a refrigerator—a fully functional refrigerator in a field with some bottles of beer inside, all nicely chilled. But how could a refrigerator be working in the middle of a field, apparently unconnected to any external energy source, yet maintaining a cold interior? And just what is it doing there, and how did it get there? You take a closer look and you see a solar panel on its top, which is connected to a battery, which in turn operates the compressor that all fridges have in order to function. So the mystery of *how* the refrigerator works is resolved. The refrigerator captures solar energy through the photovoltaic panel, so it is the sun that is the source of energy that operates the refrigerator and enables it to pump heat from cold to hot—in the opposite direction to the one that normally governs heat flow. Thus, despite Nature's drive to equalize the temperature inside and outside the cabinet, in this physical entity that we call a 'refrigerator', there exists a functional design that enables us to keep our food and drinks at a suitably low temperature.

But the mystery of how it got there in the middle of the field remains. Who put it there? And why? Now if I told you that no one put the refrigerator there—that it came about spontaneously through natural forces, you would react in total disbelief. How absurd! Impossible! Nature just doesn't operate like that! Nature doesn't spontaneously make highly organized far-from-equilibrium, purposeful entities—fridges, cars, computers, etc. Such objects are the products of human design—purposeful and deliberate. Nature, if anything, pushes systems *toward* equilibrium, toward disorder and chaos, *not* toward order and function. Or does it?

The simple truth is that the most basic living system, a bacterial cell, is a highly organized far-from-equilibrium functional system, which in a thermodynamic sense mimics the operation of a refrigerator, but is orders of magnitude more complex! The refrigerator involves the cooperative interaction of, at most, several dozen components, whereas a bacterial cell involves the interaction of thousands of different molecules and molecular aggregates, some of enormous complexity in themselves, all within a network of thousands of synchronized chemical reactions. In the case of the fridge, the function is obvious—to keep the beer or whatever else is in the cabinet cold by pumping heat from the cold interior to the hotter exterior. But what is the function of the bacterial cell with its organized complexity? Its function can be readily recognized simply by observing its action. Just as the function and workings of the refrigerator can be uncovered by inspecting its operation, so the cell's function—its purpose if you like—can be revealed by seeing what it does. And what do we see? Every living cell is effectively a highly organized factory, which, like any manmade factory, is connected to an energy source and power generator that facilitates its operation. If the energy source is cut off the factory ceases to operate. This miniature factory takes in raw material, and through the utilization of power from the factory's power generator, converts those raw materials into the many functional components, which will then be assembled to produce the factory's output. And what is that output? What does this highly elaborate nano-factory produce? More cells! Every cell is ultimately a highly organized and efficient factory for making more cells! The Nobel biologist Francois Jacob expressed it rather poetically: 'the dream of every cell, to become two cells'.

And here precisely lies the life problem. Just as the likelihood of a functional fridge—cabinet, energy collector, battery, compressor, gas—spontaneously coming together naturally seems inconceivable, even if its parts were all readily available, the likelihood for the spontaneous formation of a highly organized far-from-equilibrium miniature chemical factory—a nano-factory—also seems inconceivable. It is not just common sense that tells us that highly organized entities don't just spontaneously come about. Certain basic laws of physics preach the same sermon—systems tend toward chaos and disorder, not toward order and function. No wonder several of the great physicists of the twentieth century, amongst them Eugene Wigner, Niels Bohr, and Erwin Schrödinger, found the issue highly troublesome. Biology and physics seem contradictory, quite incompatible. No wonder the proponents of Intelligent Design manage to peddle their wares with such success!

The paradox inherent in the very existence of a living cell has profound consequences. It means that the issue of life's emergence is not just some esoteric activity of historical interest, analogous to an individual seeking to uncover his family tree. Until the paradox associated with life's emergence is resolved, we will not understand what life is. And, as final confirmation that understanding has been achieved, we will be able to translate that understanding into a coherent proposal for the synthesis of a chemical system that we would categorize as 'living'.

PROLOGUE

The purpose of this book is to reassess this enthralling subject and demonstrate that a general law that underlies the emergence, existence, and nature of all living things can now be outlined. I will argue that thanks to a newly defined area of chemistry, termed by Günter von Kiedrowski 'Systems Chemistry', the existing chasm separating chemistry and biology can now be bridged, and that *the central biological paradigm, Darwinism, is just the biological manifestation of a broader physicochemical description of natural forces.* This admittedly ambitious attempt to merge biology into chemistry rests on the idea that there is a kind of stability in nature that has been previously overlooked, one I have termed *dynamic kinetic stability*. Amalgamating that form of stability into a Darwinian view of evolution leads to a *general (or extended) theory of evolution*, encompassing both biological *and* pre-biological systems. Interestingly, Darwin himself already understood that a general principle of life is likely to exist. In a letter to George Wallich in 1882 he wrote:

> I believe that I have somewhere said (but cannot find the passage) that the principle of continuity renders it probable that the principle of life will hereafter be shown to be part, or consequence, of some general law...

This book is an attempt to demonstrate that Charles Darwin in his genius and far-sightedness was right, and that such a theory can now be formulated. I will attempt to show that chemistry, the science that bridges physics and biology, can provide answers, still in part incomplete, to these fascinating questions. Achieving a better understanding of what life is may not only tell us who and what we are, but will hopefully provide greater insight into the very nature of the cosmos and its most basic laws.

PROLOGUE

In writing this book, I have benefited from interaction with and input from many people. In particular I wish to thank Jan Engberts, Joel Harp, Sijbren Otto, and Leo Radom for detailed comments and criticisms of an early draft, to Mitchell Guss, Gerald Joyce, Elio Mattia, Elinor and David O'Neill, and Peter Strazewski for general comments, and to Gonen Ashkenasy, Stuart Kauffman, Günter von Kiedrowski, Ken Kraaijeveld, Puri Lopez-Garcia, Meir Lahav, Michael Meijler, Kepa Ruiz-Mirazo, Robert Pascal, Eörs Szathmáry, Emmanuel Tannenbaum, and Nathaniel Wagner for valuable discussions that have contributed to my understanding, and to Nella, my wife, for many discussions and for her critical eye and insights which have greatly impacted on the text. Finally I owe a very special debt to my Editor at OUP, Latha Menon. Her profound scientific understanding and remarkable editorial skills ensured the text did not stray unnecessarily into stormy biological waters and contributed greatly to its final form. Of course any errors that remain are purely my own.

1

Living Things are so Very Strange

Living and non-living entities are strikingly different, yet somehow the precise manner in which these two material forms relate to one another has remained provocatively out of reach. Life's evident design, in particular, stands out, a source of endless speculation. The creativity and precision so evident in that design is nothing less than spectacular. The structural intricacy of the eye with its iris diaphragm, the lens with its variable focal length capability, the light-sensitive retina connected to the optic nerve for information transmission, is the classic example of nature's design capability. But that's just the very tip of the design iceberg. Due to the remarkable advances in molecular biology over the past six decades we have discovered that nature's design capabilities can be immeasurably greater. Take the ribosome, for example. The ribosome is a tiny organelle present in all living cells in thousands of copies that manufactures the protein molecules on which all life is based. It effectively operates as a highly organized and intricate miniature factory, churning out those proteins—long chain-like molecules—by stitching together a hundred or more amino acid molecules in just the right order, and all within a few seconds. And this exquisitely efficient entity is contained within a complex chemical structure that is just some 20–30 nanometres in diameter—that's just 2–3 millionths of a centimetre! Think about that—an entire factory, with all the elements you'd expect to find in any regular factory, but within a structure so tiny it is completely invisible to the naked eye. Indeed, for elucidating the structure and function of this remarkable organelle, Ada Yonath from the Weizmann Institute, Israel, Venkatraman Ramakrishnan from the Laboratory of Molecular Biology at Cambridge, and Thomas Steitz from Yale University were awarded the 2009 Nobel Prize in Chemistry.

No less impressive than life's extraordinary design capabilities is its breathtaking diversity, a perpetual source of inspiration. Red roses, giraffes, butterflies, snakes, towering redwoods, whales, fungi, crocodiles, cockroaches, mosquitoes, coral reefs—the mind boggles at nature's spectacular and unmitigated creativity. Literally millions of species, and that's before we have even touched upon the hidden kingdom, the bacterial one. That invisible kingdom is itself a source of overwhelming, almost incomprehensible diversity, one that is just beginning to come to light. But life's design and diversity are just two characteristics out of a wider set that serve to compound the mystery and uniqueness of the life phenomenon. Some of life's characteristics are so striking you don't have to be too observant to notice them. Take life's independent and purposeful character, for example. You can't miss it. My granddaughter certainly didn't, even when she was just 2 years old. She clearly appreciated the distinction between a real dog and a realistically looking toy one. She happily played with toy ones, but was afraid of real ones, not being quite sure what surprise a real one might have in store for her. She learnt very quickly that a toy dog's behaviour was predictable, while a real one had a mind of its own.

But there are other characteristics of life that are less obvious at first sight, though very obvious to the scientist in the lab, which also continue to tantalize and are in need of explanation. So if we want to understand what life is, where better to begin our journey of discovery than by considering the characteristics that distinguish living things from non-living ones. Ultimately, understanding life will require us to understand those special properties, both in themselves and how they came about. Some, as we will see, may be understood in Darwinian terms, though the debate about those explanations continues. Others, however, cannot be understood that way, and their very essence continues to trouble us. They certainly troubled the great physicists of the twentieth century, amongst them Bohr, Schrödinger, and Wigner, since several of life's characteristics appear to undermine the most basic tenets of modern science. Yet other characteristics have led some modern biologists to throw up their arms in despair. How else to interpret the recent description of life by Carl Woese: 'Organisms are resilient patterns in a turbulent flow—patterns in an energy flow.'[1] That obscure remark, verging on the mystical, comes from one of the leading molecular biologists of the twentieth century—the discoverer of the Archaea, the third kingdom of life. Woese's statement reaffirms how problematic the life issue continues to be.

So we have here an intriguing phenomenon—biologists, the scientists who devote themselves to the study of living systems, and who possess a deep appreciation of life's complexity, having successfully probed many of its key components, remain mystified by what life is, and physicists, with their deep understanding of nature's most fundamental laws, are no less confused. Both continue to struggle with the nature of life question and we can only conclude that the 3,000-year 'what is life' riddle remains that—a riddle. Let us then begin our journey of discovery by briefly considering each of the characteristics that makes life special, so different to inanimate matter, and discuss what makes those characteristics so strange, so very strange.

Life's organized complexity

Living things are highly complex. In fact the very first line in Richard Dawkins classic text *The Blind Watchmaker* begins with the remark that we animals are the most complicated things in the universe.[2] That attention-grabbing line on its own is enough to drive home the realization that we animals must be something very special. But what is it about us living things that makes us so complicated, or, to use the more scientific word, so complex? And what does the term 'complex' actually mean? At the risk of sounding circular, one could say the term 'complexity' is itself complex, not readily defined, and attempts over the years to quantify the concept have not proven too successful, at least not within a biological context. Let us then focus on the crucial aspect of complexity as it pertains to biology—the highly organized nature of living things.

In the non-living world it is easy to find examples of complexity. The shape of a boulder is certainly complex and in that case the complexity derives from its irregular shape. To describe its shape with precision would require information—the more irregular the shape, the more information would be required. The physical location of each point on the boulder's surface would need to be specified in some manner. The important point, however, is that we understand that the boulder's irregularity, the source of its complexity, is *arbitrary*. It could have been any one of a zillion other irregular shapes and the boulder would still be a boulder. It is not the particular irregularity of that boulder that makes it a boulder. By contrast, in the living world complexity is not arbitrary, but highly specific. Even the slightest structural change to that organized complexity may have dramatic consequences. For example, even a single change in a human's DNA sequence, one out of 3 billion units, may potentially lead to thousands of genetic diseases, such as sickle cell anaemia, cystic fibrosis, and Huntington's disease. Small changes to life's complex structure may well undermine the viability of that living system, and in extreme cases the living system may be living no longer.

What is quite extraordinary and hard to comprehend is that such organized complexity extends to entities as small as a bacterial cell, just one thousandth of a millimetre across. In many respects the bacterial cell operates like a highly sophisticated nano-scale factory, nano-scale meaning the factory components are of molecular size, that is, of the order of one millionth of a millimetre in length. That nano-factory involves a highly complex but integrated network of chemical reactions, which extract energy from the environment, storing it in a number of different chemical forms for use in the biosynthesis of essential cellular building blocks; the control and regulation of the cellular machinery to ensure proper function; the list goes on and on. The cell is not just a master chemist, but a master physicist as well. That microscopic entity uses every mechanical trick in the tradesman's book—pumps, rotors, motors, propellers, even scissors to snip here and there, all at nano-scale, to ensure cellular functions are carried out expeditiously, as required by the cell's 'purpose'.

But that undisputed complexity, so different to inanimate complexity, is puzzling and raises two immediate questions. How is the organized complexity of the cell maintained, and how did it come into being? Organized complexity and one of the most fundamental laws of the universe—the Second Law of Thermodynamics—are inherently adversarial. We won't go into the Second Law in any detail at this stage, but a very simple (and limited) expression of the Second Law is the statement that organized systems spontaneously tend toward disorganization, toward disorder. Nature prefers chaos to order, so disorganization is the natural order. Take a pack of cards in some highly ordered sequence—say four aces, followed by four kings, then by four queens, and so on, down to four twos—shuffle the deck and the sequence invariably becomes disordered. You'll almost certainly end up with some random sequence. The likelihood of some other highly ordered sequence being formed is very slight. That's the Second Law in action. The state of my desk at any point in time is further proof, if it were needed. No matter how often I tidy my desk, it always seems to quickly revert to its preferred disorganized state. Within living systems, however, the highly organized state that is absolutely essential for viable biological function is somehow maintained with remarkable precision. There is even a biological term for the phenomenon whereby that organized state is maintained—*homeostasis*, from the Greek meaning 'standing still'.

So how is the cell's organized complexity maintained, if a central law of physics and chemistry is constantly operating to undermine it? The answer to this first question is relatively easy, at least within the context of the Second Law: the living cell is able to maintain its structural integrity and its organization through the continual utilization of energy, which is in fact part of the cell's *modus operandi*. That's why we have to eat regularly to survive—to furnish the body with the necessary energy to enable the body's regulatory mechanisms to maintain life's organized homeostatic state. That also explains how my desk gets to be tidy occasionally—I expend energy now and then to restore a semblance of order whenever my desk has become too disordered to be functional. So there is no thermodynamic contradiction in life's organized high-energy state, just as there is no contradiction in a car being able to drive uphill in opposition to the Earth's gravitational pull, or a refrigerator in maintaining a cool interior despite the constant flow of heat into that interior from the warmer exterior. Both the car driving uphill and the refrigerator with its cold interior can maintain their energetically unstable state through the continual utilization of energy. In the car's case the burning of gasoline in the car's engine is the energy source, while in the case of the refrigerator, the energy source is the electricity supply that operates the refrigerator's compressor. In an analogous manner, energetically speaking, the body can maintain its highly organized state through the continual utilization of energy from some external source—the chemical energy inherent within the foods we eat, or, in the case of plants, the solar energy that is captured by the chlorophyll pigment found in all plants. No fundamental problem there.

But how the initial organization associated with the simplest living system came about originally is a much tougher question. Despite the widespread view that Darwinian evolution has been able to explain the emergence of biological complexity, that is not the case. Darwinian evolution *is* able to broadly explain how a simple single-cell living organism—what one might call the microbial Adam—eventually became an elephant, a whale, or a human. But Darwinian theory does not deal with the question how that primordial living thing was able to come into being. The troublesome question still in search of an answer is: *how did a system capable of evolving come about in the first place?* Darwinian theory is a *biological* theory and therefore deals with *biological* systems, whereas the origin of life problem is a *chemical* problem, and chemical problems are best solved with chemical (and physical) theories. Attempting to explain chemical phenomena with biological concepts is methodologically problematic for reasons we will discuss subsequently, and in some sense that approach may have been partly responsible for the conceptual dead-end the subject seems to have found itself in.

Significantly, Darwin himself explicitly avoided the origin of life question, recognizing that within the existing state of knowledge the question was premature, that its resolution at that time was out of reach. So the question of how the first microscopic complexity came into being remains problematic and highly contentious. Did a cellular precursor to that exquisitely complex miniature factory that is the living cell come together purely by chance, by the various bits and pieces randomly linking up in precisely the right manner? Not very likely. To draw on an analogy popularized by Fred Hoyle, the well-known astronomer (though famously misapplied), the likelihood of such an event would be similar to that of a whirlwind blowing through a junkyard and assembling a Boeing 747. Life's organized complexity *is* strange, very strange. And how it came about is even stranger.

Life's purposeful character

There is another facet to the organized complexity of living systems that has been strikingly evident to humankind for thousands of years—life's purposeful character. That purposeful character is so well defined and unambiguous that biologists have come up with a special name for it—teleonomy. The 'teleonomy' word was introduced about half a century ago to distinguish it from the 'teleology' word with its cosmic implications, and we will have more to say about how these terms relate to one another in chapters 2 and 8. At this point let us simply note that teleonomy, as a biological phenomenon, is empirically irrefutable. The term simply gives a name to a pattern of behaviour that is unambiguous—all living things behave as if they have an agenda. Every living thing goes about its business of living—building nests, collecting food, protecting the young, and, of course, reproducing. In fact, within the biological world that's how we broadly understand and predict what goes on. We understand a mother nurturing her offspring. We know better (or should know better) than to step between a mother bear and her cub. We understand two males competing for a female; we understand a stray cat rummaging through a trash bin. We intuitively understand the operation of the biological world, including, of course, all human activity, through life's teleonomic character.

In the non-living world, by comparison, understanding and prediction are achieved on the basis of quite different principles. No teleonomy there, just the established laws of physics and chemistry. You throw a ball into the air and you want to know where it will land? The precise landing point is not calculated by considering the ball's purpose. The ball has no purpose. Only Newton's laws of motion will provide the answer. You mix some chemical compounds together and you want to know whether they will react and what materials are likely to form? You consider and apply the appropriate chemical rules, depending on the nature of the problem, and you come up with a prediction. No purpose, no agenda—just inviolate laws of nature. The notion of purpose within the inanimate world was laid to rest with the modern scientific revolution of the seventeenth century.

The very existence of teleonomy however, leads us to a strange, even weird, reality: in some fundamental sense we are simultaneously living in *two* worlds each governed by its own set of rules—the laws of physics and chemistry within the inanimate world and the teleonomic principle that dominates the biological world. Indeed, given the existence of two distinct worlds we find ourselves interacting quite differently with each of those worlds. Consider our interactions within the inanimate world. We move from one place to another as required, we try to keep warm when it is cold, to keep dry when it rains, we build a physical enclosure to live in to protect ourselves and to facilitate life's activities. We learn to climb up slopes despite the gravitational force, to generate fire for cooking, to manufacture tools for improved function, to plug a hole in a leaking roof, to avoid physical injury, and so on. All of our interactions with the inanimate world are based on the recognition that there are certain laws of nature, described primarily by the physical sciences, which govern the manner in which the universe functions.

Understanding those laws helps us to keep out of trouble, and, even better, enables us to take advantage of nature's *modus operandi*, thereby allowing us to further life's goals more effectively. In fact that is the essence of technology—creating systems that exploit nature's laws in a beneficial manner.

Our interactions with the living world, however, are of a quite different kind and are much more complex. As we have already noted, the living world is teleonomic—all living creatures are busy furthering their agenda, and in doing so they must take into account the particular agenda of other living beings. Accordingly, living things create a web of interaction with other living things, making many of our actions mutually dependent. Consider us humans. We communicate and deal with members of our immediate family, with our work colleagues, with other members of our society in an endless series of interactions—by spoken and written word, more subtly without words, by gestures. Some of these interactions are cooperative in nature, some competitive. Ordering a cappuccino at the local café or going to the hairdresser exemplify cooperative interactions, while bargaining in the market over the price of some article or fending off an intruder are competitive interactions. Our lives involve endless interactions of both types as we individually pursue our 'purpose' and get on with life's goals. We also continually interact with a wide range of non-human life forms. Our need for sustenance is satisfied by feeding on other living creatures, both animal and vegetable, and we protect ourselves against the life forms that threaten us, whether multicellular creatures—bears, sharks, snakes, mosquitoes, or spiders—or from single-celled creatures—bacteria of endless variety. Many non-human interactions are cooperative—the pet dog that we feed which provides companionship and warns us of intruders, the billions of bacteria in our gut to which we happily provide room and board, and who return the favour by assisting us with our digestion and more.

We are so used to this dual state of affairs—matter that exists in both living and non-living forms—that much of what has been said here is glaringly obvious and very much taken for granted. Familiarity breeds acceptance, if not contempt. But if I were to tell you that on Mars all material forms obeyed one set of principles, yet on Venus they followed another different set, we would all be startled. How could that be? Two material forms broadly following two distinct sets of principles? The fact that here on Earth there exist two material forms that are distinct in character, are governed by different organizational principles, which comfortably coexist, and in fact continually undergo material interchange—non-living matter is continually transformed into living matter, and vice versa—demands some explanation. How can this stark duality in the nature of matter exist and what does it signify?

Before going any further let me be unequivocal and make one point perfectly clear: it goes without saying that within the teleonomic world the same underlying rules of physics and chemistry that govern the inanimate world are still operative. No doubt about that. When a person falls off a ladder the law of gravity is operative in exactly the same way as when a bag of sugar falls off a shelf. But in many respects those natural laws are of little or no use when applied to living systems. The law of gravity and the Second Law of Thermodynamics aren't particularly helpful when you are arguing with a neighbour over some property issue, or when seeking to renew an expired licence, or when fending off an aggressive dog. Within the living world those same laws have little predictive value—they are certainly operative but appear to be of only secondary importance. The underlying rules of physics and chemistry have somehow been taken hostage and overwhelmed by another more dominant set of principles. If you want to predict the actions of a crouching lion preparing to pounce on an unsuspecting zebra, a mother tending to her young, a lawyer planning to sue you on behalf of an aggrieved client, or indeed any other teleonomic action, the laws of physics and chemistry are of little use. Neither a physicist nor a chemist will be able to offer a useful prediction. If you want to make a prediction about some impending event in the living world, go ask a biologist, psychologist, economist, lawyer, or other teleonomic specialist, depending on the nature of the question.

Not surprisingly then, much of human knowledge and understanding involves the teleonomic, rather than the physicochemical world. Consider for a moment any large university with its many faculties, each dedicated to a particular field of enquiry. The faculties of humanities, commerce, and law (and to a lesser extent, the faculty of medicine), are dedicated to the teleonomic world with its many manifestations. There is just one faculty—the faculty of natural sciences—that dedicates itself specifically to the study of the natural world, and even within this faculty we find the department of biological sciences grappling awkwardly with the teleonomic reality, uncertain as to how the paradox of a dichotomic world can and should be resolved. That, then, is the undeniable, yet so far inexplicable reality—the laws of nature, as primarily articulated in the subjects of physics and chemistry, offer few insights into the predominantly teleonomic world of which we find ourselves very much a part.

LIVING THINGS ARE SO VERY STRANGE

Intriguingly, despite the irrefutable teleonomic character of living systems, some biologists still have difficulty in coming to terms with that extraordinary character. The troublesome 'purpose' word, now sanitized and repackaged into the scientifically acceptable 'teleonomy' word, still leaves many modern biologists squirming uncomfortably. The scientific revolution's overthrow of 2,000 years of teleological thinking has left biologists anxious and unwilling to accept even the slightest vestige of that earlier, misplaced way of thinking. But there is no denying the teleonomic principle. The evidence supporting it is simply overwhelming, all around, literally endless, and cannot simply be dismissed out of hand.

In fact, it is intriguing to point out that those biologists who have argued against the concept of teleonomy, have, without realizing it, demonstrated their total faith in the principle by their everyday actions. Those scientists, like us all, actually stake their lives on its validity. Every time we get into a motor car, for example, we are betting our lives on teleonomy! Our purpose in getting into our car is to get to some destination, and to do so safely. On the roads we have to manoeuvre through an endless stream of vehicular metal—the other cars—careering about hither and yon, a real threat to life and limb. The consequences of a collision between any two metal hunks can be personally disastrous, yet we happily accept that risk day by day. Why? Because of teleonomy. We know that within every other metal hunk careering about, there is a driver whose purpose is identical to our own—to get to his destination in one piece! Though one occasionally comes across an erratic driver who seems to prove the exception to the teleonomic rule, for most of us, on most days, that teleonomic principle operates reliably and, as anticipated, we arrive at our destination safely. So those so-called disbelievers in teleonomy are actually silent and committed believers. The world we have to navigate our way through on a daily basis is composed of both biological and non-biological systems. When dealing with the non-biological world we intuitively apply the laws of physics and chemistry. But, consciously or unconsciously, no person would be able to get through a single day without continuous application of the teleonomic principle. No doubt whatever, in the living world, teleonomy, as a predictive and explanatory principle, is the way to go.

The fact that multicellular animals, like us, behave in a purposeful manner may not appear that surprising. After all, as already noted, we animals are highly complex—we possess a brain and nervous system so it might be argued that in us animals the teleonomic character is just a reflection of significant neural complexity. But here's the surprise. It is not just multicellular cognitive beings—humans, monkeys, camels, and the like, with a brain and central nervous system that manifest this teleonomic character. That character is also clearly manifest at the level of the single cell! Put a bacterium in a glucose solution in which the glucose concentration is variable and the bacterium 'swims' toward the high concentration region. That phenomenon is called chemotaxis. The bacterium, which utilizes the glucose's chemical energy to power its metabolic processes, is effectively going out for dinner, much like the crouching lion about to pounce on a zebra.

Of course a bacterial cell cannot swim in the conventional sense of the word. A simple bacterium such as *E. Coli* is powered by several flagella, which, depending on the direction of flagella rotation, enable the bacterium to direct its motion within the solution. If the solution contains nutrition, then the bacterium rotates the flagella in one direction such that its motion is toward the nutrition. However, if the solution contains toxins, then it rotates the flagella in the opposite direction causing the bacterium to tumble, thereby changing its direction away from those toxins. The directed swimming action of the bacterium is unambiguous: without a brain or in fact any neural activity whatever, that clump of chemical aggregates within a membrane (which is itself a chemical aggregate) that we call a living bacterium follows its agenda of seeking out its next meal, keeping out of trouble, and getting on with its life. The fundamental behavioural patterns of bacteria and humans are not as different as one might initially conceive.

We have focused on the teleonomic behaviour of the living cell but in point of fact it's not just the *actions* of the bacterium that reflect its teleonomic character. The highly complex cell *structure* that we have already discussed is the most explicit and profound expression of that teleonomic character. Pretty well every element within that bacterium can be associated with a particular cell function, in much the same way that the individual components of a clock—pendulum, cogs and wheels, springs, hands, cabinet, etc.—can also be associated with a particular function, except that within the cell the structural complexity and intricacy is orders of magnitude greater. The long string of Nobel prizes awarded to the pioneers of cell structure and function, beginning with Watson and Crick's 1953 structure elucidation of DNA, the molecule of heredity, attests to the importance that the scientific community has attributed to these landmark discoveries. The simplest cell is a marvel of teleonomic design, breathtaking in its intricacy and efficiency. Bottom line: teleonomy is as evident at the single-cell level as at the multicell level. The living world screams out teleonomy no matter where you look.

Our confident recognition that teleonomy is an undeniable and legitimate concept does raise a problem. We believe in a material world, we no longer believe in a vital force, we now know living things are made up of the same 'dead' molecules as non-living ones, so what is going on here? How can *any* organization of inert matter come to life? How can *any* natural organization of matter act on its own behalf? How is it that a small crystal of sugar, say, about the same size as that bacterial cell that we've been discussing, behaves so differently from the bacterial cell? It is true that the sugar crystal is composed of just one single organic compound, sucrose, whereas the bacterial cell is made up of thousands of different organic molecules and molecular aggregates enclosed within a membrane. But how can this complex mixture of organic materials behave so differently from the single organic compound, sucrose? Just mixing together a thousand different organic materials in any and every combination certainly does not create a living system.

What then is the nature and source of life's apparent *élan vital*, that teleonomic character already evident in a bacterial cell? How is it possible for the living world to be seemingly governed by different laws from those that are operational in the inanimate world? If we wish to understand life, we will need to provide a rationale for life's teleonomic character in the same chemical terms we use to explain the global characteristics of inanimate systems. Simply sweeping the issue of teleonomy under the complexity carpet with a shallow explanation of 'emergent properties of complex systems' will not suffice. Such a response is little more than dressing up the discredited *élan vital* concept in scientifically more acceptable attire.

As we will discuss in chapter 2, Jacques Monod, the French biologist, who won the Nobel Prize for his contribution to the understanding of DNA replication and its role in protein synthesis, and who had a deep appreciation of the complex chemical behaviour of the living cell, was confounded by this apparent paradox. No wonder then that the great physicists of the twentieth century were both fascinated and troubled by this duality in material behaviour. The question of teleonomy is one with profound scientific and philosophical implications. If we ultimately believe in the material nature of living things, then life's teleonomic character should somehow be a manifestation of the material that produces that teleonomic character, just as the hardness of a crystal of salt or the softness of a rubber ball are understandable characteristics of the materials from which these objects are made. We won't understand life till we understand teleonomy. Indeed, as part of our goal of understanding life, in chapter 8, I will propose a physicochemical characterization of teleonomy, as well as a mechanism for its emergence.

We have noted that biological systems are purposeful both in form and action. But what exactly *is* the purpose? Can it be specified? Ask a number of different people what their goal or purpose in life is, and you'll get a variety of answers. One person might say their goal is to travel the world, another to make a lot of money, yet another to make the national Olympic team, to get married and have ten kids, yet another to write a book on the nature of life. The list is endless. Of course any one person might have a number of different goals in mind. We humans are a restless species, never entirely satisfied. But if we want to get to the very essence of biological purpose, we need to get away from multicellular complex beings and look at the simplest life form, that simple cell, the prokaryotic (without a nucleus) bacterium. As we have seen, everything that the single bacterial cell does, every aspect of its highly complex internal structure, is teleonomic, and the entire teleonomic apparatus associated with that bacterial cell is directed toward one goal—cell division. In recognizing that fact for single-cell creatures, one can discern that in multicellular creatures that replicating drive is also immensely powerful. Ultimately many of the life goals of living creatures, if not explicitly related to reproduction, can be understood as indirectly related to reproduction, as a means to that end. Living things, even the very simplest ones, *are* strange, yes, very strange.

One final comment concerning the reality of teleonomy and whether it can serve as a totally legitimate scientific concept. The argument might be put that teleonomy is only conceptual, merely in our minds, not real like physical forces, such as gravity. However, this distinction is not as valid as it might initially seem. True, teleonomy is conceptual, it *is* just in our minds. Teleonomy is indeed a construct, intangible in a physical sense, one that enables us to better understand the biological world. But now think about the Newtonian concept of gravity for a moment. That's a real force, right? But what does 'real' actually mean? Have you ever seen, heard, or touched a gravitational field? Is there some sophisticated scientific instrument that will reveal such a field, say by capturing its image? The answer is no. A gravitational field is not directly observable in any way—it also is a concept, just like the teleonomic principle. It is useful to talk about gravitational fields because the concept enables us to explain the behaviour of matter—falling apples, for example. Metaphysically speaking, however, both gravity and teleonomy are mental constructs that assist us in organizing the world around us. Inductive reasoning, which we will discuss in chapter 3, is, in its very essence, conceptual. All inferred patterns are conceptual and are nowhere to be found except within our minds. True, the concept of gravity can be quantified, while the teleonomic concept cannot, and quantifiable concepts, quite rightly, have a preferred status in science compared to non-quantifiable ones. But the fact that a concept is not quantifiable does not make it any less real than one that is. If we are all willing on a daily basis to get into our cars and stake our lives on the validity of the teleonomic principle, then, despite it not being quantifiable, we must all be quite convinced of its reality.

Life's dynamic character

We have discussed in some detail the fact that the living cell is a highly organized entity and compared it to a familiar mechanical entity, a clock. Both are organized in the sense that all of the component parts contribute to the operation of the holistic entity. The parts of the clock enable it to fulfil its function of telling the time, the parts of the cell enable it to fulfil its function and become two cells. Of course the clock is an organized entity that has been constructed to fulfil its particular function—it is man-made, whereas the bacterial cell has somehow come about of its own accord. Nevertheless, the machine metaphor for understanding living systems has been useful and allowed us to continue to probe cell function, to discover in ever greater detail the precise workings of this remarkable 'machine'. Closer examination of the two 'machines', however, reveals an extraordinary distinction between the machine-like characteristics of the clock and the cell.

Within the clock the components remain in place and continue to operate until one or other of them wears out and the system ceases to function. But within the living cell the situation is spectacularly different. Whereas a clock is a static system, whose parts are permanent and unchanged, every living system is *dynamic*. Its parts are continually being turned over. Let me explain.

You meet an old friend that you haven't seen in a few years and you greet him with the comment: 'Hi Bill, great to see you again, you haven't changed a bit!' You make that comment because Bill looks very much as you remember him from your last encounter. But here is an extraordinary fact. The person standing in front of you, who looks like Bill, talks like Bill, and is called Bill, is, materially speaking, effectively a totally different person from the Bill you saw some time back. Just about every molecule in Bill's body has been replaced since you last saw him. Almost all the stuff of which Bill (and you and me) is made has been turned over. For some parts of us, our hair and fingernails, for example, that turnover is obvious. But for the rest of what makes you, you, the turnover is hidden from view. It takes place surreptitiously. Like all human beings you are primarily composed of the some 10 thousand billion (10,000,000,000,000) cells that make up your body. (We actually also contain within our bodies some 100 thousand billion foreign cells, bacteria, but we'll get to the significance of that later in the book.) And each of those cells is itself composed of an array of biomolecules—lipids, proteins, nucleic acids, and so on.

Consider proteins, as they are the archetypal molecules of life. Much of life's infrastructure is based on the huge array of different proteins in our body. Muscle is protein, cartilage is protein, enzymes are proteins, indeed much of the internal workings of the cell chemistry involve protein molecules. And now to the key point: due to the critical importance of proteins in governing life's functions, protein structure must be strictly regulated and controlled to ensure no damaging mutations have taken place in their structure over time. The consequences of such mutations could well be catastrophic—even cell death. Protein integrity is crucial for life's successful function. Several years ago a key mechanism for maintaining the proteins' structural integrity was discovered by Avram Hershko, Aaron Ciechanover, two researchers at the Technion in Israel, and Irwin Rose, at the University of California at Irvine, for which they received the Nobel Prize in chemistry in 2004. What they discovered was that intracellular protein is continually being turned over—cellular protein is constantly being degraded and resynthesized in a tightly regulated process.

At least one of the reasons for that dynamic process is to ensure that the proteins' structural integrity is maintained. The mechanism of that process need not concern us here, but the net effect of this protein regulation and control mechanism is that even within a few hours much of the cellular protein in our bodies has been degraded and reconstituted. And if that dynamic molecular character isn't enough to get you wondering, let's also point out that at the *cellular* level the degree of turnover is no less impressive. Your blood cells, billions of them, are replaced daily, your skin cells continually turn over. In fact in an adult human being hundreds of billions of new cells are created daily and these new cells are created in order to replace a similar number that die, many by design, through what is termed programmed cell death.

The bottom line: essentially all of the stuff that makes you, you, and Bill, Bill, is being constantly turned over so that over a period of weeks you are in a strictly material sense a totally different person. The 'life as a machine' analogy, though of value, offers no hint of life's underlying dynamic character. Yes, life *is* very strange. Answering the 'what is life' question will have to come up with a good explanation for life's dynamic and ephemeral nature.

Life's diversity

As we have already commented, life is spectacularly diverse. True, there is considerable diversity in the non-living world, but the diversity of the living world is quite different in character. Non-living diversity is arbitrary, while living diversity seems deliberate, coherent. Look at the plant kingdom, look at the animal kingdom—literally millions of different species, each perfectly adapted to function and survive in its particular ecological niche. Life's staggering and very special diversity in all its grandeur is out there, everywhere, all around us.

But the macroscopic diversity that we see around us is just the tip of the diversity iceberg. The largely invisible microbial world is where the concept of diversity takes on new meaning. Microbes are effectively everywhere. An early estimate of the earth's bacterial biomass puts it at 2×10^{14} tons.[3] That's sufficient to cover the earth's land surface to a depth of 1.5 metres! More recently it has been discovered that a litre of sea water can contain as many as one billion bacteria[4] emphasizing how little we know about that invisible world. Indeed, estimating bacterial diversity is still in its infancy due to the difficulties in culturing and sequencing diverse microbial populations. By some estimates the number of bacterial species in a gram of soil could be in the order of a million and a common estimate of the number of all bacterial species on earth range between 10 million and one billion. Let me be clear here—we are speaking of the number of bacterial species, not the number of bacteria! In fact the diversity is so overwhelming that microbial genomicists have started to think in terms of 'species genomes' or *pangenomes* that possess a common core of shared genes. Individual genomes are too diverse to allow meaningful characterization. What is clear and beyond dispute is that the diversity in the microbial world is one of staggering proportions.

What is remarkable, however, is that the underlying basis for life's diversity continues to trouble biologists, beginning with Charles Darwin and through to the present day. In his *Origin of Species* text, Darwin proposed a Principle of Divergence, though from that monumental work it is not entirely clear whether the Principle of Divergence derives from his primary principle, the Principle of Natural Selection, or should be considered as an independent principle. Darwin himself seemed ambivalent on this point. The source of the conflict is clear: divergence means that *many are derived from few*, whereas selection (of any kind, natural or otherwise) means *many are reduced to few*. The two are inherently contradictory and no amount of verbal gymnastics can get around that. No wonder then that attempts to reconcile the apparently irreconcilable continues to torment modern biologists.[5,6] What is clear is that the source of life's diversity does begin with reproductive variation, though the detailed manner in which that variation leads to speciation and diversity remains controversial. In chapter 8, we will propose a physical approach to the problem of diversity in the living world and the cooperative nature of biological interaction that has accompanied that diversity.

Life's far-from-equilibrium state

Earlier we discussed how the emergence of life's organized complexity constitutes a thermodynamic puzzle. But there is another facet of life's nature that is related to that complexity, which is also troubling with respect to the Second Law of Thermodynamics—its far-from-equilibrium state. Consider a bird that is hovering in space, maintaining an almost stationary position by flapping its wings. Clearly that bird is in an unstable state. If it were to stop flapping its wings, it would drop to the ground. However, that bird is able to maintain its unstable state, suspended in mid-air by the continual expenditure of energy. By constantly flapping its wings the bird is essentially pushing down on the air, and so is able to overcome the earth's gravitational pull.

The example of the hovering bird and its unstable state might seem to be a transient moment, of no general significance. But from a purely energetic point of view the hovering bird's unstable state is actually a metaphor for all living things. Consider the energetics of the simplest life form, a bacterial cell. That cell, from a thermodynamic point of view, is also unstable and exists in what is termed a far-from-equilibrium state in that it also must continuously expend energy to maintain that state. There are many aspects to that far-from-equilibrium state but to illustrate the point we will just describe one—the existence and maintenance of ion concentration gradients in living cells. Let us describe what that means. You dissolve some table salt, sodium chloride with the chemical formula NaCl, in water, and what happens is that the crystals of salt break up into their two constituent ions, the sodium ion, Na^+, and the chloride ion, Cl⁻. Initially the concentrations of the two ions in solution would not be uniform, but would be higher near the point of dissolution. After some time, however, the ions would, by diffusion, distribute themselves evenly throughout the solution. That is, yet again, an example of the operation of the Second Law. A situation where there is a high concentration in one part of the solution and a low concentration in another part would be unstable compared to a uniform distribution and the Second Law is quick to correct this non-uniform ion distribution.

For living cells, however, inherently unstable ion concentration gradients are *essential* for many physiological functions so a non-uniform ion distribution, termed an ion concentration gradient, exists between the cell's interior and its exterior, *despite* the Second Law, and that gradient is maintained over time. How can that be? In order to maintain inherently unstable concentration gradients over time the cell has to operate ion pumps, pumping ions *against* the gradient—just like the bird flapping its wings to stay aloft. Of course, in order to operate those ion pumps, the cell must utilize energy, and that energy has to be supplied to the cell in some form, as discussed earlier.

In other words, there is no thermodynamic mystery in the *ability* of cells to maintain that far-from-equilibrium state—they can do so by the continual expenditure of energy that is constantly supplied by the environment. However, there *is* a deep mystery hidden in the scheme we've just described, even if thermodynamically speaking the energy book-keeping has been meticulously maintained. *Just how could far-from-equilibrium chemical systems have come about in the first place?* If, as we believe, chemical processes led to the emergence of life on earth, how could chemical processes on the prebiotic earth that would be driven *toward* their equilibrium state, meaning toward chemical systems of low energy, have led to the emergence of *complex, high-energy, far-from-equilibrium* systems? Recall, the Second Law states that all systems seek to become *more* stable, yet in the process of emergence exactly the opposite must have taken place. In the context of the Second Law the emergence of unstable, far-from-equilibrium systems might be paraphrased: *you can't get there from here*. But we did! The troubling question is then how did we?

Life's chiral nature

Many of the molecules found in living systems are chiral, meaning that the molecule's mirror image is not superimposable on the molecule itself. Our two hands reflect that quality—a left hand is the mirror image of a right hand, but the two hands are not superimposable on one another (Fig. 1). The term 'handedness' is in fact a commonly used metaphor to express this characteristic of chirality in a molecule, and in order to distinguish between these two chiral forms, different classifications are possible. One of the earlier ones, still prevalent in biology today, is the D, L classification, where one chiral molecule is labelled D (for dextro, or right-handed) and its mirror image, L (for levo, or left-handed), based on its spatial relationship to the organic substance, glyceraldehyde. The point is that the physical and chemical properties of two chiral molecules, D and L, are identical (though there are some exceptions that we need not concern ourselves with here). That also suggests that in an arbitrary environment the two chiral molecules should be present in equal amounts. If, however, for whatever reason we start off with a quantity of some chiral material of a single chirality, say all D, then

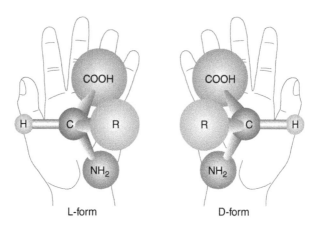

Fig. 1. Handedness associated with chiral objects. An object is chiral if its mirror image is not superimposable on itself.

that same Second Law of Thermodynamics discussed earlier, tells us, that given enough time, that material of single chirality will become racemic, meaning that the material will end up consisting of equal quantities of D and L forms (due to slow D to L and L to D interconversion). Simply, a racemic mixture is more stable than a single chiral form—it is more disordered, and therefore will tend to be established given enough time.

We commenced this topic with the statement that many of the molecules of life are chiral. The amino acid building blocks from which all proteins are constructed, and sugars, from which nucleic acids and carbohydrates are composed, are all chiral. What is important, however, is that within living systems only one chiral form of the two possible chiral forms is present—biological sugars are almost invariably D-sugars, while amino acids are almost invariably L-amino acids. Living systems are universally *homochiral* (meaning of just one chirality). But this homochirality raises two fundamental questions.

First, how did the homochirality of life emerge in the first place? Given the chiral nature of many objects in the world, how did homochirality of living things come about from a world that is intrinsically heterochiral, or, put differently, how did a world with its inherent *two-handedness* become *single-handed* within its biological part? And, second, once homochiral systems did emerge by some means, how can its maintenance be explained, given that heterochirality (an equal mixture of two chiral forms) is inherently more stable than homochirality? In that sense the homochiral nature of life represents yet another manifestation of life's unstable and far-from-equilibrium character described earlier.

* * *

The above detailed description of living states and their unique characteristics should serve as a stark reminder how strikingly different living and non-living systems actually are. Actually that in itself would not be a problem. Within the inanimate world different material forms can also express very different properties. Some are solid, some liquid, some gases, some conduct electricity, some don't. Some are coloured, some are colourless. But these differences are readily explained by basic chemical theory. Consider, for example, the three traditional states of matter of water—ice, liquid water, steam. The first is a brittle crystalline solid, the second a clear colourless liquid, and the third an invisible gas—you can't get much more different than that! But despite the dramatically different properties of the three states, we fully 'understand' those three states of matter. No mystery, no confusion.

So what is the basis for that understanding? Our understanding is based on our molecular view of matter and the associated kinetic theory which tell us that the states of matter depend on the magnitude of the forces operating between the individual molecules. The stronger those intermolecular forces, the more likely the substance will be solid. Of course the temperature of the material also has a bearing on the state of matter that is obtained. The higher the temperature, the more likely the material will be gaseous, due to the higher kinetic energy of the individual molecules. Thus the particular properties of ice, water, steam, derive directly from our molecular view of matter; the physical sciences have provided us with a pattern that enables us to convincingly relate the three states of matter to each other. Most significantly, the final and definitive confirmation that we do indeed 'understand' the three states of matter comes about through our ability to readily convert one state to another. Indeed, as predicted by what are termed phase diagrams, we can bring about those transformations in different ways. We can convert ice to water by either applying pressure or by heating, and we are able to convert ice to steam without having to pass through the water phase. In summary then, we say we 'understand' the three states of matter, solid, liquid, gas, because we can (a) explain the different properties of those different states in fundamental molecular terms, and (b) most importantly, our understanding provides us with control over the system in question—we know different ways of converting one state to another.

With respect to the biological world, however, our current understanding of material systems is unable to address life's unique characteristics that we've discussed in some detail. Simply put, within the material world there exists an entire class of material systems—the biological class—that exhibits a distinct pattern of behaviour that remains unexplained in chemical terms. And, paradoxically, that lack of understanding accompanies us *despite* the fact that the intricate mechanisms of biological function *are* increasingly understood. Somehow we know more and more of the cell's mechanisms, yet that molecular knowledge seems to bring us no closer to understanding the essence of biological reality. We see lots and lots of trees, but a view of the forest remains frustratingly obscure. Understanding life will require that we are able to offer unambiguous explanations for life's unique characteristics. That is one key challenge this book will attempt to address.

2

The Quest for a Theory of Life

In the previous chapter, we highlighted life's puzzling characteristics and described our inability to explain those characteristics in simple chemical terms. Not surprisingly, given the fundamental nature of the problem, attempts to understand life have weighed upon humankind for several millennia, so let us briefly review the central concepts that have moulded our thinking through the ages. Aristotle's ideas, going back over 2,000 years, have been particularly influential as they stemmed directly from his extensive studies of living things—Aristotle was a dedicated biologist both in practice and in spirit. That detailed observation of living things was responsible for what might be considered his most important contribution to scientific thought—his *teleological* view of nature, a view of such powerful persuasion that it ended up dominating Western thinking for over two millennia.

Simply put, Aristotle saw in the processes by which life is generated and maintained one that indicates them to be *goal directed*. Every aspect of reproduction and embryonic development, for example, exemplifies that purposeful and goal-directed character. Given that purpose was so clearly associated with such a wide variety of material forms (though all examples came from the biological world), it only seemed logical to conclude that an underlying purpose was associated with *all* material forms, biological *and* non-biological (Aristotle's famous Final Cause). Indeed, that is the essence of Aristotle's teleological view—that there is an underlying purpose to the workings of nature, that purpose governs the cosmos as a whole. Given the bountiful biological evidence for Aristotle's teleological argument, in retrospect it is quite understandable that teleological thinking held up largely uncontested for over two millennia.

But then in the sixteenth century the beginnings of an intellectual stirring took place which before too long built up into a tsunami, an intellectual storm that transformed the scientific landscape of the time. What is now termed the modern scientific revolution, whose central figures include Copernicus, Descartes, Galileo, Newton, and Bacon, radically changed mankind's perception of the universe and his proper place in it. Its major accomplishment: the long-standing teleological view of the universe underwent a dramatic reassessment and, in scientific quarters at least, was effectively discarded. In what was a dramatic turnaround from those 2,000 years of deeply entrenched and established thinking, that revolution dismissed the idea of an underlying purpose in nature, and replaced it by a view—indeed, the very essence of the modern scientific revolution—that *nature is objective*, that there is no underlying purpose to the natural order. The scientific and philosophic implications of that revolution cannot be overstated. Jacques Monod, in fact, considers that idea the *single most important idea* offered by man over the 150,000–200,000 years that he has inhabited the planet. That single idea propelled mankind into a new conceptual reality, one whose ultimate significance and impact we have yet to fully discover. But, paradoxically, that revolutionary idea, together with the accompanying change in man's perception of the universe, only served to raise serious difficulties with regard to the life issue. Indeed the change in scientific perception ended up *accentuating* the life riddle by the creation of what appeared to be undeniable contradictions within the *new* scientific thinking. Prior to the modern scientific revolution a unity of sorts could be found in man's view of the cosmos; teleology encompassed both the animate *and* inanimate worlds. But as a direct result of that revolution, the need to explain the existence of two worlds, and the nature of the relationship between those two worlds, necessarily arose. Remarkably then, the modern scientific revolution was not only unable to satisfy mankind's relentless urge to find his proper place in the universe, but placed new and seemingly greater obstacles along the path to an improved understanding of the material world, a world that necessarily incorporates both animate and inanimate.

The next major step in this ongoing saga was the 1859 landmark publication of Charles Darwin's *On the Origin of Species*. Remarkably, Darwin's theory of evolution, though offering a grand unification of biology, only served to widen the chasm separating animate and inanimate. As previously mentioned, the scientific revolution of the seventeenth century was slow in coming about because Aristotle's teleological argument was so persuasive, so logical, so empirically based—the world around us simply exudes endless examples of purposeful design, though, of course, that entire edifice of purpose rests on a biological foundation. In his paradigm-shattering thesis,

Darwin swept away the most compelling basis for believing in a teleological universe by the profound insight that a simple mechanistic explanation—natural selection—lay behind the emergence of purposeful design in living systems. Through his principle of natural selection, Darwin was able to extend and reinforce the scientific revolution, a revolution based on the axiomatic premiss of an objective universe, into that one area where it had seemed awkwardly inapplicable—into biology. Following that epoch-making contribution, cosmic teleology, at least in scientific circles, was finally laid to rest.

However, though Darwin did provide a 'physical' explanation as to how simple life evolved into increasingly complex life, Darwin did not explain, or even attempt to explain, the manner by which inanimate matter was transformed into simple life. Interestingly, that problematic omission was already obvious during Darwin's time, notably by Darwin himself. In a letter to a botanist colleague he remarked: 'it is mere rubbish thinking at present of the origin of life; one might as well think of the origin of matter'. Darwin deliberately side-stepped the challenge, recognizing that it could not be adequately addressed within the existing state of knowledge. Ernst Haeckel, one of Darwin's contemporaries, put it rather less kindly with his comment: 'the chief defect of the Darwinian theory is that it throws no light on the origin of the primitive organism—probably a simple cell—from which all the others have descended. When Darwin assumes a special creative act for this first species, he is not consistent, and, I think, not quite sincere...'[7] The central question of how life emerged—how design, function, and purpose were generated and incorporated into *non-living* matter, remained unresolved, a perpetual thorn in the side of the physical sciences.

The dramatic advances in physics that took place in the first decades of the twentieth century failed in their turn to clarify the issue. Indeed, in 1933, Niels Bohr, one of the fathers of atomic theory, in a famous 'Light and Life' lecture, went as far as to propose 'that life is consistent with, but undecidable or unknowable by human reasoning from physics and chemistry'.[8] Effectively, Bohr extended what he perceived as the 'irrationality' of quantum theory, one that physicists were forced to accept and accommodate, to biological systems as well. A kind of intrinsic biological irrationality! Living and non-living things can exist in two kinds of material form, and that is that. Erwin Schrödinger, the father of quantum mechanics, whose provocative little book, *What is Life?*,[9] we mentioned earlier, was particularly puzzled by life's strange thermodynamic behaviour. Simply, modern physics and biology appeared quite at odds—fundamentally incompatible. Schrödinger found himself following Bohr's line of reasoning, and concluded, rather enigmatically, that living matter, while not eluding the established laws of physics, was likely to involve 'other laws of physics' hitherto unknown.

A generation later Jacques Monod, the Nobel biologist, in his classic 1971 monograph *Chance and Necessity*,[10] lucidly reaffirmed the existence of a deep physics–biology divide, a divide only widened by the scientific revolution. The main issue that troubled Monod was life's teleonomic nature. The very existence of that teleonomic character appeared to violate one of the fundamental principles of modern science—the objectivity of nature. Monod summarized the problem as follows:

> Here therefore, at least in appearance, lies a profound epistemological contradiction. In fact the central problem of biology lies

with this very contradiction, which, if it is only apparent, must be resolved; or else proven to be utterly insoluble, if that should indeed turn out to be the case.

Simply put, how could function and purpose have emerged from an objective universe devoid of function and purpose? So though Aristotelian teleology had been vanquished by the new scientific order, its elimination left a troublesome vacuum. The scientific reality of teleonomy, so evident in every facet of the biological world, was undeniable. No cosmic implications there, just down-to-earth biological empiricism. But what is the source of this teleonomic character? How could purpose of *any* kind emerge from an objective universe? The conclusion seems inescapable: understanding life will require that we understand teleonomy—the two are necessarily and inexorably linked. But there is a positive aspect to this analysis. If we are able to explain the physical basis of teleonomy, it might provide mechanistic insight into the means by which life itself emerged. We will argue for such a connection in chapters 7 and 8.

In retrospect one might be tempted to say that part of the difficulty that physicists, such as Bohr and Schrödinger, had in addressing the life problem lay with the fact that the problem of what is life and how it emerged is fundamentally a *chemical* problem. After all, both the processes that govern the function of living systems, as well as the ones that presumably led to the emergence of living systems from inanimate matter, primarily take place at the scientific level of enquiry we call chemistry. But if one might consider that ignorance with regard to the chemical mechanisms of life was the missing element needed to properly address Schrödinger's question, the dramatic developments within molecular biology over the half-century following Schrödinger's work proved otherwise. Watson and Crick's 1953 landmark DNA study[11] signalled the beginnings of a true revolution in our understanding of the cell-based machinery, the machinery of life. Major discoveries quickly followed—the mechanisms of DNA replication, protein synthesis, energy transduction, and central metabolic cycles, to name just a few. Truly dramatic advances in our understanding of many of the molecular mechanisms of life took place in rapid succession. Yet, paradoxically, our digging deeper and deeper into the mechanisms of life did not seem to lead us any closer to being able to address Schrödinger's basic 'what is life' question, or the related question—how did life emerge? In fact, in 1974, twenty years after the discovery of DNA, Karl Popper, the iconic philosopher of science, supported the Bohr–Schrödinger view with his assertion that the origin of life problem was 'an impenetrable barrier to science and a residue to all attempts to reduce biology to chemistry and physics'.[12] And the very same Francis Crick of DNA fame, in a 1981 text, *Life Itself* considered the emergence of life so miraculous an event that he even entertained the possibility of 'directed panspermia', the extreme idea that life on earth originated from outer space by the deliberate seeding of the earth by some alien life form![13]

The conclusion is quite striking. In the broadest sense we have made surprisingly little progress regarding the 'what is life' question since Charles Darwin. Yes, we now know that all life is cell based, that genetic information is coded in the DNA molecule, that the proteins of life so critical to all of life's functionality are expressed through a universal code that relates the DNA sequence to particular amino acids, that there is a universal energy storage facility based on the ATP molecule. But that detailed molecular understanding, of enormous significance in its own right, has only served to substantiate Darwin's original claim—that all life is derived from some early common ancestor, that life is one thing. Darwin, of course, was lacking the plethora of mechanistic details that modern molecular biology has generously bestowed on us, but the belief in the unity of life, the insight that all life is related through physical law, was the essence of his contribution and the basis of the Darwinian revolution. Quite remarkably then, the molecular insights showered upon us by sixty years of extraordinary discoveries in molecular biology do not seem to have brought us any closer to resolving the 'what is life' question. Yes, as we have already noted, we can see many, many trees in the forest of life, but the view of the forest itself remains frustratingly obscure.

Defining life

Enormous effort has gone into attempts to define life over the years and we will end this section by considering some of the more recent ones. That brief survey will only serve to reaffirm how confused the life topic has become. Literally hundreds of definitions have been proposed over the years and there are few signs that the flow is abating. In *Searching for the Definition and Origin of Life*,[14] Radu Popa lists forty definitions that were proposed in 2002 alone, the last full year before his book was published, suggesting that the process of defining life has within it streaks of autocatalytic character. And therein lies the problem—the plethora of different definitions of life, many incompatible, if not outright contradictory, make it clear there is some inherent difficulty with the 'definition of life' endeavour. Stepping back and reflecting on this expanding literature from a distance brings to mind the metaphor of a dog chasing its tail. Let's consider several recent examples of life definitions arbitrarily chosen from Popa's list to illustrate the problem first hand.

> Life is defined as a material system that can acquire, store, process, and use information to organize its activities.[15]
> Life is defined as a system of nucleic acid and protein polymerases with a constant supply of monomers, energy and protection.[16]
> Life is defined as a system capable of 1. self-organization; 2. self-replication; 3. evolution through mutation; 4. metabolism; and 5. concentrative encapsulation.[17]
> Life is simply a particular state of organized instability.[18]

The above definitions, all relatively recent and all insightful in their own way, show almost no overlap. If all of the definitions hadn't begun with the two words 'life is...', we would be excused for believing that these definitions were about totally different concepts. The first, by Freeman Dyson, focuses on information (software); the second, by Victor Kunin, on the nucleic acid and protein infrastructure (hardware) and the energy required to drive the process; the third, by Gustaf Arrhenius, attempts to specify several of the characteristics that living things share; while the fourth, by Remy Hennet, addresses life's thermodynamic aspect. And had we been willing to list other definitions from the many others on offer, we would have been able to come up with more definitional variety. Life is indeed many things, yet none alone is life.

Finally let us consider the most common and generally accepted definition of life, the one proposed within the NASA Exobiology Program in 1992, and generally referred to as the NASA definition of life: *Life is a self-sustained chemical system capable of undergoing Darwinian evolution*. Though attractive in some respects, it also suffers from certain deficiencies. The first might be considered a technical one. The NASA definition could be understood to refer to *individual* life forms, say, a bacterium, an elephant, or a human. However individual life forms cannot undergo evolution; they can only reproduce and die. It is only *populations* of living things that are able to undergo Darwinian evolution. But even ignoring that technical aspect, the definition remains problematic as it has obvious exceptions. A mule, the offspring from the mating of a horse and a donkey, is sterile, so it clearly cannot reproduce. That of course means that a population of mules cannot undergo Darwinian evolution, even though we all agree that mules are alive. The same goes for solitary rabbits—unable to reproduce, yet very much alive. This criticism, based on mules and single rabbits, has been expressed quite frequently in recent years and through repetition seems to have lost some of its force. However familiarity should in no way undermine its relevance and validity. The criticism is soundly based and cannot be ignored. Like so many life definitions, it is too easy to cite exceptions. Invariably living things are either *excluded* from the various definitions or non-living things are improperly *included* in them.

So how to proceed? In an insightful article published a decade ago, Carol Cleland, who teaches philosophy at the University of Colorado, and Christopher Chyba, a Princeton University astronomer, changed the very nature of the debate.[19] They pointed out that attempting to define life before we understand what life is, is to put the cart before the horse. Seeking the definition of an entity that we *do* understand is problematic enough. Attempting to define an entity that we are still struggling to understand is futile. Based on the Cleland and Chyba argument, we can now identify the fundamental problem with the NASA definition. The NASA definition does not attempt to tell us what life *is*, but rather how we might recognize it. Just as water's physical characteristics might help us determine if some liquid is water or not, the NASA definition may be able to inform us if something is alive by seeing whether it does something that living things typically do (undergo Darwinian evolution). Cleland and Chyba claim that what is needed is not a definition of life, but a comprehensive *theory of life*. We will describe our attempts in that direction in the final two chapters.

To sum up, this brief historical survey has illustrated the confusion that the life issue has generated over the centuries right through to the present day, as well as some of the reasons that the long-standing 'what is life' riddle has remained unresolved. Until the deep conceptual chasm that continues to separate living and non-living is bridged, until the two sciences—physics and biology—can merge naturally, the nature of life, and hence man's place in the universe, will continue to remain gnawingly uncertain.

3

Understanding 'Understanding'

The previous chapter indicated that we are still lacking a theory of life, a theory that will enable us to understand what life is and how it emerged, that despite the recent detailed insights into life's mechanism, something central is missing in our understanding of the life phenomenon. But what exactly do we mean by the term 'understand'? When addressing most day-to-day questions, there seems to be no need to explain the term—it is self-evident. But when addressing the life question, the issue turns out to be more complex. What we mean by 'understanding' goes to the very heart of the scientific method and beyond, forcing us to at least briefly address basic philosophical questions that have weighed on mankind for over 2,000 years.

In the scientific world we strive to achieve understanding of phenomena in the world around us through application of the *scientific method*. The method is well known so we will just address those aspects that will be relevant to our analysis. At the very heart of the scientific method is the *process of induction*, a way of reasoning whose roots can be traced back to ancient Greek philosophy, but was raised to scientific prominence with its formal description by Francis Bacon, one of the fathers of the modern scientific revolution. This may all sound quite formal, even esoteric. But the essence of the methodology is actually very simple. So simple in fact that even young children intuitively understand it and (unconsciously) apply it quite routinely. Indeed, I would argue that the essence of all scientific endeavour, stripped of its many elaborations, trimmings, and jargon, is nothing more than the successful application of the inductive method. It is the successful application of the inductive method that forms the basis for what we term 'understanding'.

Inductive reasoning involves the reaching of general conclusions from a set of empirically obtained facts—what one might simplistically term *pattern recognition*. Consider a very simple example: the falling of apples. Indeed without exception, all apples do fall, so one can reasonably formulate a general rule of nature: 'apples fall'. However, even the less observant amongst us will have noticed that it is not just apples that fall, but that all material objects display that same falling characteristic. Accordingly, the limited 'apples fall' rule can be further extended to an 'all objects fall' rule, though the behaviour of certain objects, such as hot-air balloons, requires the pattern to be elaborated further to account for these apparent exceptions.

Needless to say the phenomenon of falling objects is so obvious that even a small child grasps its essence very quickly and in doing so has applied the inductive method at a fundamental level. When a child drops some object and it falls to the ground, it doesn't take too long before the child 'understands' that the singular event of the falling object manifests the general 'objects fall' rule. So even young children, with no knowledge of induction or the scientific method, intuitively apply the principles of induction to better understand and adapt to the world around them. Thomas Macaulay, a British poet and historian, pointed this out already over 150 years ago with his comment:

> The inductive method has been practised ever since the beginning of the world by every human being. It is constantly practised by the most ignorant clown, by the most thoughtless schoolboy, by the very child at the breast. That method leads the clown to the conclusion that if he sows barley he shall not reap wheat. By that method the schoolboy learns that a cloudy day is the best for catching trout. The very infant, we imagine, is led by induction to expect milk from his mother or nurse, and none from his father.[20]

In fact *all* cognitive beings, human and non-humans alike, apply the method routinely, whether consciously or subconsciously, in a process that has been deeply engrained in us all by evolution. Yes, your pet dog, despite his lack of familiarity with Bacon's treatise, or epistemology in general, also routinely applies the inductive method. Just watch his reaction when you begin to open a can of his favourite dog food. Based on the pattern he has learnt to recognize over time, he fully understands that he is about to get fed. It is that evolutionarily acquired ability to gather empirical information and to recognize patterns within that gathered information which provides cognitive beings with the ability to respond to the external world in a beneficial manner (from the point of view of the cognitive being). Both your dog, a 2-year-old child, and the scientist in the lab are applying the same inductive methodology, the difference only being in the level of sophistication of the patterns that are recognized.

As mentioned above, small children recognize the 'objects fall' rule. But it took the genius of an Isaac Newton to recognize a much broader pattern, one which links the behaviour of falling apples to the orbits of celestial bodies, such as the moon and the earth—a law of gravity that describes the interaction of physical bodies in precise mathematical terms. So when we say we understand *why* apples fall and *why* the moon rotates around the earth, it is because both these specific events exemplify a more general pattern, one that governs the behaviour of all physical bodies. But what that means, however, is that there is no *absolute and deep* understanding as to *why* apples fall. Gravity is just the name of the general pattern to which the falling apple event belongs.

Ultimately *all* scientific explanations are inductive—they involve no more than the recognition of patterns and the association of the specific within the general. Broadly speaking the wider the generalization, i.e., the greater the number of empirical observations that are embraced by the generalization, the greater its predictive power and the more significant the generalization. Simplistically, that's what modern physics is all about—seeking ever-general laws that underlie the workings of the universe, extending the pattern. So that is what Einstein's special and general theories of relativity do—they extend and generalize the more limited Newtonian pattern. With his theory of relativity Einstein was able to place Newton's gravitational force in a more general context, and in that sense it constituted an advance on the Newtonian description.

According to Einstein, gravity is just the natural movement of objects through curved four-dimensional spacetime, thereby providing a more general basis for understanding a wide range of physical phenomena, including the behaviour of falling apples. And, of course, physicists are still at it, attempting to further generalize, with sophisticated formulations such as string theory and M-theory, constantly working toward the so-called final theory—the theory of everything, the ultimate pattern. Of course whether an ultimate pattern is achievable is another question, one that belongs within the realms of philosophy, not just science—a wonderful question in its own right, but one that goes well beyond the scope of this discussion.

The role of mathematics in generating patterns is crucially important. The ability to express the pattern quantitatively through the language of mathematics greatly enhances the predictive power of the generalization and therefore its utility. Richard Feynman, the Nobel physicist, once compared the accuracy of quantum theories to the ability to measure the width of North America to an accuracy of one hair's breadth. Now that's a pattern we should take note of! Such predictive capabilities ensure that mathematics plays a central role in pattern formulation, though this is not to dismiss the value and utility of qualitative patterns. Let us not forget the revolutionary impact of Darwin's ideas of natural selection and common descent, ideas that were entirely qualitative in their formulation yet continue to profoundly impact on man's view of himself to this very day. To quote the aphorism attributed to Albert Einstein: *Not everything that counts can be counted, and not everything that can be counted, counts.*

We have used the term 'patterns' to describe what it is that the inductive method seeks, though scientists typically use other terms, such as hypotheses, theories, laws, to mention the main ones, the difference being primarily in the degree to which the pattern has been confirmed. Thus Newton's Law of Gravity is uncontroversially considered to be a law due to the innumerable times apples and other objects have fallen, and the regularity with which the sun rises every day. However, the term 'pattern' with its inherent fuzziness, does have its advantages. In contrast to terms such as 'theories' and 'laws' which radiate some sense of absolute truth, the term 'pattern' is more subtle, less committed, less definitive, more open to modification. Even Newton's laws, those pertaining to gravity and motion, have had to undergo revision following Einstein's revolutionary insights. If we keep in mind that every hypothesis, theory, or law is ultimately just a pattern, the day that theory or law is modified or revoked will be less surprising, less disconcerting.

As to the underlying reason for the existence of those patterns, rules, laws, generalizations, or whatever we wish to call them, science is unable and does not pretend to address such questions. Despite the widespread view that the laws of nature are the explanation of natural phenomena, Ludwig Wittgenstein, the great twentieth-century philosopher, pointed out almost a century ago in his famous *Tractatus* (Latin for *treatise*) that 'the whole modern conception of the world is founded on the illusion that the so-called laws of nature are the explanations of natural phenomena.' There is no fundamental explanation for *any* phenomenon and the best we can do is to say that the pattern is the explanation. Patterns are the link between the underlying reality and our understanding of that reality. The basis for the patterns, those underlying laws of nature, are fascinating questions in their own right, but these are philosophical questions, beyond the strict scientific domain, and therefore outside the scope of this discussion. To quote Wittgenstein yet again: 'whereof one cannot speak, thereof one must be silent'.

Given the above statements it can be appreciated that there are degrees to understanding, that understanding is to a significant extent *subjective*, because the process of pattern recognition is not always definitive. Pattern recognition is, to some extent, in the eye of the beholder. As the Nobel physicist Steven Weinberg lucidly pointed out, as good a way as any to establish whether a pattern is insightful is to see whether it induces an 'Aha!' from colleagues. Having said that, however, it is clear that the nature of understanding within physics, a more fundamental science, is quite different from its operation within biology, whose domain is the study of inherently highly complex systems. Within physics generalizations are invariably rigorously quantified, articulated in the language of mathematics so that exceptions to the rule are not tolerated and require a reformulation of that rule. Within biology generalizations are frequently qualitative and exceptions to the rule are not just tolerated, but accepted as normal. In any case, regardless of the field of endeavour, it should be emphasized that the same set of observations may on occasion be interpreted in different ways and so may lead to the recognition of different patterns.

This is particularly true when the observed patterns are statistical rather than absolute, as is common in the social sciences, or when the patterns are qualitative rather than quantitative in nature. It is for this very reason that historians frequently come up with quite different models for understanding a set of historic events, since those events may be successfully organized in more than one pattern. The extensive literature on the causes of the First World War exemplifies the way an unambiguous set of historical events can be understood and interpreted in different ways. Nor do patterns have to be mutually exclusive. Both a 2-year-old child and a theoretical physicist have some understanding of why apples fall, though their explanations differ markedly. Both see in the falling apple the manifestation of a more general pattern, though the physicist recognizes a pattern that is both broader and quantifiable. Significantly however, the child's simple 'falling object' rule is sufficient to serve him extremely well on a day-to-day basis. So provided that the child has no immediate plans to launch a satellite into space or undertake space travel, then for all practical purposes the extra insight that Newton's law of gravity and Einstein's theories of relativity offer into the behaviour of matter, beyond that offered by the 'objects fall' rule, will be of little consequence. In fact, if one thinks about it, the physicist about to undertake some mountain climbing is most likely to be applying the 'objects fall' rule to guide him in his adventure, rather than string theory or special and general theories of relativity.

In conclusion, when a system can be patterned in more than one way, the question as to which pattern is better may well depend on the particular application. The title of a 2009 Woody Allen movie, *Whatever Works*, captures the essential idea. Yes, that sums it up nicely—whatever works. Ultimately, whatever one calls them—theories, laws, models, hypotheses, patterns—all efforts to find order in our universe can never fully capture the reality of nature. The patterns we uncover are merely *reflections* of that reality—some better, some worse, whose recognition brings us some sense of order to the complex world that we find ourselves in. The preceding discussion will now assist us in addressing a central issue in the continuing search for biological understanding—the issue of reduction versus holism.

Reduction or holism

We pointed out earlier that the inductive method—the seeking of generalizations, the recognition of patterns—is at the core of all scientific understanding. However, a particular kind of inductive thinking has proven to be of special value, the one termed *reduction*. The concept of reduction can itself be elaborated upon and split up into a number of subgroups, something philosophers of science have been exploring in recent years, but these more detailed ideas need not concern us here. The essence of the reductionist approach is simply: 'the whole can be understood in terms of the interaction of its constituent parts'. For example, if you want to understand how a clock works then break it up into its component parts—wheels, cogs, springs, etc., and see how these work together to create the functional entity. Reductionist thinking of one kind or other has been instrumental in advancing scientific understanding from the earliest days of the scientific revolution.

In opposition to the reductionist view is a more recent school of thought termed *holism*, whose philosophy can be summarized by the simple statement: 'the whole is more than the sum of its parts', and so appears to negate the reductionist view. Holism contends that within complex systems in particular, unexpected emergent properties arise that cannot be derived by examining the individual components of the system (by emergent properties we mean that there are properties at the higher and more complex level that are not observed at lower levels). This approach has gained considerable influence in recent years, specifically with regard to the biological sciences, due to the extraordinary complexity of even so-called 'simple' biological systems, and has led to the establishment of a new branch in biology—systems biology. Carl Woese's view of biological systems as 'complex dynamic organization', rather than as a 'molecular machine' whose behaviour can be understood from its component parts, illustrates this new 'systems' way of thinking.[1]

So which is the better approach for addressing biological problems—reduction or holism? That depends on who you ask. Jacques Monod[10] offered a rather disparaging view of holism (and holists) with his comment: 'A most foolish and wrongheaded quarrel it is, merely testifying to the "holists" [sic] profound misappreciation of the scientific method and of the crucial role analysis plays in it.' The confusion surrounding the apparent conflict between reductionism and holism as applied to biological systems is a long-standing one and graphically illustrated in the proceedings of a conference entitled 'Problems of Reduction in Biology' attended by a group of leading biologists and philosophers, including Peter Medawar, Jacques Monod, and Karl Popper that took place in September 1972, in Bellagio, Italy. At the end of that meeting June Goodfield was reported as saying:

> I am overpowered by a feeling of *déjà vu* verging at times on the very edge of intellectual impotence. 'Reductionism'; 'anti-reductionism'; 'beyond reductionism'; 'holism'.... The issue is a very old one recurring in various forms with unfailing regularity throughout biological history, and the feeling of impotence arises because, after all this time, the issue never seems to get any clearer.[21]

Well, almost forty years on and little seems to have changed. Reduction and holism in biology seem as controversial now as then. A recent polemical essay by Denis Noble that comes down firmly on the side of holism, discusses the same dilemmas, though illustrated with examples from modern systems biology.[22] Carl Woese, a reborn holist, puts it even more starkly:

> Biology today is at a crossroad. The molecular paradigm, which so successfully guided the discipline throughout most of the 20th century, is no longer a reliable guide. Its vision of biology now realized, the molecular paradigm has run its course. Biology, therefore, has a choice to make, between the comfortable path of continuing to follow molecular biology's lead or the more invigorating one of seeking a new and inspiring vision of the living world, one that addresses the major problems in biology that 20th century

biology, molecular biology, could not handle and, so, avoided. The former course, though highly productive, is certain to turn biology into an engineering discipline. The latter holds the promise of making biology an even more fundamental science, one that, along with physics, probes and defines the nature of reality.[1]

Powerful and provocative words indeed. But in a sharp critique of holism, the Nobel biologist, Sydney Brenner, recently wrote: 'The new science of systems biology claims to be able to solve the problem but I contend that this approach will fail because deducing models of function from the behaviour of a complex system is an inverse problem that is impossible to solve.'[23]

Despite that treacherously uncertain backdrop, let us now briefly venture into this philosophic lion's den. I will offer some thoughts on this troublesome philosophic divide and how it impacts on our goal of better understanding living systems. At least in the context of life, I propose that the reductionist–holistic divide is more semantic than substantive, and that holism, when probed more deeply, can be thought of as just a more elaborate form of reduction.

At the risk of gross oversimplification we may state that the most useful application of reductionist philosophy, when viewed as a scientific methodology, is the one termed 'hierarchical reduction', the idea being that phenomena at one hierarchical level can be explained using concepts taken from a lower hierarchical level. Steven Weinberg recently expressed the idea succinctly: 'explanatory arrows always point downward'.[24] Thus, to illustrate, one attempts to explain social behaviour based on individual organismic behaviour, organismic behaviour in terms of cellular behaviour, cellular behaviour based on biochemical cycles, and biochemical cycles rest upon more basic physical and chemical concepts of molecular structure and reactivity, and so on, continuing down to fundamental subatomic particles. Hierarchical reduction seeks to provide understanding level by level, with phenomena at each level being explained by the conceptual framework associated with the level immediately below. Much of the spectacular advance witnessed in the physical sciences since the scientific revolution of the seventeenth century can be directly attributed to the successful implementation of that methodology. Within the biological sciences, the reductionist harvest has been particularly abundant. The enormous advances in our understanding of biological processes, such as DNA replication, protein synthesis, metabolic cycles, etc., all derive from the reductionist methodology. Without question molecular biology has revealed many of the wonders of cell function at the molecular level—reduction *par excellence*.

But, as noted in chapter 1, the enormous complexity of biological systems often makes the reductionist methodology difficult to implement, and it is that difficulty that has been responsible for the burgeoning anti-reductionist, holistic approach to biological systems of recent decades. The holistic view derives its persuasive influence from the *systems theory* school of thought that builds on the idea that within complex systems, systemic relations arise that produce novel and quite unpredictable characteristics. So, in recalling Weinberg's reductionist comment 'explanatory arrows always point downward', together with June Goodfield's despairing commentary, how are we to respond to the two opposing viewpoints? And what are the implications of this apparently fundamental disagreement with respect to our attempts to understand life?

To a large extent criticism of the reductionist approach derives from extreme expressions of reduction, such as the one offered by Francis

Crick,[25] who claimed that 'the ultimate aim of the modern movement in biology is to explain all biology in terms of physics and chemistry'. Such claims appear unrealistic for the foreseeable future, and remain an ultimate aim in just the same way that the 'ultimate aim' of chemistry is to predict all chemical phenomena through solving Schrödinger's famous wave equation. In that sense the broadly based critique of reduction, given its inherent limitations, is on solid ground. But the idea that a more measured reductionist approach is unable to deal with emergent properties at all is clearly incorrect; emergent properties are regularly addressed and understood through reduction.

To take a simple example, consider the physical properties of condensed states (that's just the term for solids and liquids) that we discussed earlier. Condensed states exhibit a variety of emergent properties that are totally absent at the single molecule level. The condensed state may be solid or liquid, it may be conducting or insulating, shiny or dull. A single molecule does not possess any of those condensed state properties. A single molecule is neither solid, nor liquid, neither shiny nor dull. Nonetheless, despite the absence of these collective properties at the molecular level, these condensed state properties are well understood based on the electronic characteristics of the *individual* molecules. So we understand why at room temperature molecular hydrogen is a gas, water is a liquid, and regular table salt is a solid, based solely on properties of the individual molecules (molecular weight, charge character, etc.) and the corresponding intermolecular forces that would be expected in those materials. Similarly we may usefully predict the solid state conductivity of a material by carrying out a particular kind of theoretical analysis on the *individual* isolated molecule.

My point is that physics and chemistry are replete with such reductionist analyses that offer insight into the underlying reasons for a wide range of emergent properties. The oft cited claim that some properties cannot be explained by reduction because they are emergent is simply incorrect, though, of course, this does not mean that *all* emergent properties can be explained by reduction. Reduction as a methodology does have its limitations, as does any methodology. Complex systems cannot always be readily reduced to their component parts. Unexpected emergent properties can and do appear and in those cases, it could be argued, a holistic approach may be necessary. But a deeper appraisal of the holistic view suggests that its anti-reductionist claim is misstated to a degree. The problem lies primarily with the meaning that the term 'holistic' conveys. If 'holistic' is intended to convey the impression that the entire system is treated as a whole entity, that reduction into components is avoided, then that is certainly not the case. The systems approach dissects the complex whole into component parts as does the reductionist approach, but addresses the complex nature of interactions within the system in a more realistic fashion. The holistic view recognizes that in addition to 'upward causation' from lower-level hierarchies to higher ones, one must also consider the possibility of 'downward causation' where higher-level phenomena influence actions at lower levels.

These kinds of feedback effects can lead to quite unexpected emergent properties that cannot be easily foreseen and are not readily amenable to a simple reductionist analysis. Nonetheless a moment's thought reveals that a reductionist philosophy is at the heart of holism as well. The holistic systems approach to understanding the complexity of a biological system continues to reduce the complex system into simpler elements, though placing greater emphasis on the complex nature of the interactions between those elements. In other words the holistic approach merely preaches a more elaborate form of reduction, one that recognizes that causal relations within a system can be more complex than those implied by a simple bottom-up causal chain. To quote Athel Cornish-Bowden, the British biologist:

> the classical reductionist approach to science can be understood as a way of understanding the functioning of a whole system in terms of the properties of its parts, but now we must learn to understand the parts in terms of the whole.[26]

Reduction as an explanatory tool in science is difficult to circumvent because reduction is a key means of obtaining scientific understanding. Despite several decades of groping expectantly toward some kind of non-reductionist or even anti-reductionist methodology, that activity does not seem as yet to have born edible fruit. Holism, despite its name, can be thought of as just a reductionist elaboration, a potentially valuable elaboration for sure, but an elaboration nonetheless. Reduction in its various forms and subforms, was, is, and will likely remain the central conceptual tool in scientific endeavour. To the extent that the 'what is life' question *can* be satisfactorily resolved, I believe it can only be through a fundamentally reductionist approach—by seeking the underlying connections between chemistry and biology, by identifying the process responsible for biological complexification. Ultimately the difference between animate and inanimate must be reduced to differences in the nature of the materials within the two worlds and, in particular, in the way those materials interact and react.

4

Stability and Instability

Why do chemical reactions occur?

All living things involve chemical reactions, thousands of them, and the living cell, the basic unit comprising all life, is a highly complex set of these reactions somehow integrated into a coordinated whole. This fact alone makes the problem of understanding the living state of matter and the elucidation of its underlying characteristics a difficult one. How can that complex interplay of reactions and the molecular entities on which they operate be unravelled? Are some reactions central while others are peripheral? Of course, if we are seeking a better understanding of the reactions of life, we first need to understand chemical reactions in general. What *is* a chemical reaction and why do they take place? So let us begin by making some general comments about chemical reactivity. The subject is complex, one that requires textbook coverage for a proper treatment. Here I will give a greatly simplified version that primarily addresses those aspects of reactivity that we will need for our subsequent analysis. Our analysis will reveal that there *is* something very special within the set of chemical reactions that constitute life and understanding what that special feature is will be a focus of the ensuing chapters.

All chemical reactions involve the transformation of some chemical material into some other material. The neutralization of an acid by a base, the degradation of a protein into its constituent amino acid building blocks, the explosive reaction of a mixture of hydrogen and oxygen gases to give water, are all examples of common chemical reactions. This last reaction, that of hydrogen and oxygen gases, occurs very readily—a spark or the presence of a catalyst (for example, metallic platinum or palladium) is all that is needed for it to take place. The reverse reaction in which water spontaneously breaks up into hydrogen and oxygen gases does not occur. Why is that? What governs the direction of a chemical reaction? Broadly speaking, the answer is given by a central law of chemistry, one we have already met briefly—the Second Law of Thermodynamics.

The Second Law is actually a fundamental law of physics, so its wide applicability means that it has a number of different formulations. But in the present context it will suffice to say that chemical reactions proceed such that *less stable* materials are transformed into *more stable* materials. A ball rolling down a slope is a useful analogy. Chemical reactions proceed in a 'downhill direction', where downhill signifies toward more stable products, products that are characterized by what is termed lower 'free energy'. Since the free energy of water is lower than the free energy of a mixture of hydrogen and oxygen gases, the two gases react to form water, and the energy that was stored in the higher-energy hydrogen and oxygen molecules is released as heat. The reverse reaction in which water would be

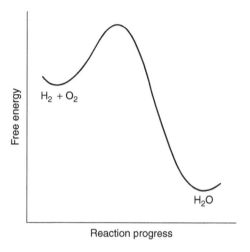

Fig. 2. Diagram illustrating the free energy change for the reaction of hydrogen and oxygen gases ($H_2 + O_2$) to give water (H_2O).

transformed into hydrogen and oxygen gases cannot take place spontaneously because that would be equivalent to a ball rolling uphill.

The relative free energies of a hydrogen and oxygen mixture compared with that of water are shown schematically in Fig. 2. The hydrogen and oxygen molecules on the left side of the diagram ($H_2 + O_2$) are located at higher energy than the water product (H_2O) on the right side of the diagram.

The diagram also reveals another important point—the hydrogen and oxygen reactants are separated from the water product by a barrier. Even though the hydrogen and oxygen gas mixture is higher in free energy than water, the path leading from reactants to products does not go downhill smoothly. It climbs uphill to some extent before it begins to descend, which means that before the reaction can proceed, the barrier must first be overcome. That's why a spark or catalyst is needed to get the reaction going. The spark provides the initial energy boost in order to get the reactants over the barrier, after which the downhill trajectory of the reaction profile takes care of the rest. A catalyst may obviate the need for a spark by reducing the barrier height so that no activation is needed and the reaction can proceed without that energy boost.

Two important lessons can be learnt from the above example. First, reactions will only take place if the reaction products are of lower free energy than the reactants. That determines the direction of any chemical reaction and is called the thermodynamic consideration. Accordingly, the Second Law of Thermodynamics indicates beforehand which reactions are possible and which are not. Once a reaction mixture has reached the lowest possible free energy state for that particular combination of materials, the system is said to be at equilibrium and no further reaction will take place. Like balls at the bottom of a valley, they have nowhere lower to roll. But the fact that a reaction mixture is not at equilibrium, i.e., not in that lowest possible free energy state, does not mean it will necessarily react. If that reaction system is trapped in a local minimum, that is, behind a barrier, it may not be able to overcome the barrier that separates that local minimum from the deeper, product minimum, much like a ball that is trapped in a hollow halfway down some slope. That's why hydrogen and oxygen gases may be mixed without any reaction taking place if neither catalyst nor spark are provided. These simple notions can now be expressed in the language of chemistry: a reaction that *is* allowed thermodynamically may or may not proceed, depending on kinetic factors (the barrier height). However, a reaction that is forbidden thermodynamically *cannot* proceed.

Entropy and the Second Law

We have seen that chemical reactions will only proceed if they are in accord with the Second Law. But it will help subsequent discussion to introduce another important concept—entropy. Understanding entropy is important because it is a key component of stability and, in fact, the Second Law can be expressed entirely in terms of entropy.

Entropy can be thought of intuitively as the degree of disorder in a system. If you throw a number of building blocks onto a surface, they are likely to fall into a disorganized pile rather than to stack up in an ordered manner. The tendency to disorder is inherent in the Second Law—ordered systems tend toward disorder, and this can be explained in statistical terms. Chemical systems respond to the drive toward disorder in exactly the same way and for exactly the same reasons as do tidy desks. Regardless of energy considerations, a chemical reaction that combines two species into one is *unfavourable* from an entropic point of view since that *increases* the order of the system (i.e. decreases its entropy), while a reaction that breaks up a single molecule into several fragments is *favoured* entropically as it *decreases* the order (increases the entropy) of the system. Accordingly, the free energy of a system incorporates within it an entropic contribution.

Replication and molecular replicators

Catalysts are frequently involved in chemical reactions. In fact, one could confidently say that almost any chemical reaction can be

STABILITY AND INSTABILITY

catalysed by some appropriate material. Within biological systems catalysts play a crucial role and are called enzymes. Without the appropriate enzyme(s) most biological reactions would either proceed very slowly, or not at all. Normally the product of a reaction and the catalyst for that reaction are different materials. In the above example of hydrogen and oxygen reacting to give water, the product is water and the catalyst would be some metal or metallic compound. But consider a reaction in which the product and the catalyst are one and the same, i.e., the product acts as a catalyst in its own formation. Such a reaction is termed *autocatalytic* for obvious reasons—the catalyst catalyses its *own* formation, rather than the formation of some other material. At first glance catalysis and autocatalysis may not seem too different. But a simple calculation of the rates at which the two reactions proceed reveals how spectacularly wrong that initial impression is. If one starts each of the two reactions, catalysis and autocatalysis, with just one *single* molecule of catalyst (or autocatalyst), a simple calculation reveals that the time required to make a small amount of material (say 100 grams) by each pathway is dramatically different. For the catalytic reaction the calculated time frame comes out in *billions of years*. For the autocatalytic reaction the corresponding calculated time frame works out at a *tiny fraction of a second*! A comparison of two seemingly similar processes doesn't get more different than that. (It should be stated that the difference between the two numbers was spectacularly large because we started off in each case with just one molecule of reactant, but even with larger quantities of starting material the effect remains dramatic.) Let me jump way ahead for a moment and state that the essence of life will be found to lie in the dramatic difference between the rates of catalytic and autocatalytic reactions. But we have quite a way to go in this discussion before the basis for that statement becomes clear.

How can that dramatic difference in reaction rate between catalysis and autocatalysis be explained? Simply put—the power of exponentials. The difference comes about because in the autocatalytic reaction, the rate of product formation proceeds *exponentially*, whereas in the catalytic reaction the rate of production proceeds *linearly*, and that difference could not be more profound. If that sounds too mathematical, let's explain the difference by recounting the classical legend of the Chinese emperor who was saved in battle by a peasant farmer. When the emperor asked the farmer how he could reward him, the farmer took out a standard chess board and asked that he be rewarded with a quantity of rice, and that the required quantity be established by a simple formula—placing a single grain of rice on the first square, two grains on the second square, four on the third, and so on, right through to the 64th square. The request sounded absurdly modest and the emperor was surprised that the peasant would be happy with such a small reward. After all, how much rice could be needed? Half a sack, a whole sack? But the truth is that the amount of rice needed to comply with the peasant's request is spectacularly large. Mathematically the total number of grains of rice placed on the board would be $2^{64} - 1$. That works out at close to 2×10^{19} grains—that's a lot of rice; more than could be found in the emperor's cellars, as well as in all the world's Chinese restaurants, and, in fact, more than exists anywhere on the entire planet. That quantity of rice, if it existed, would cover the entire earth's surface to a depth of several centimetres.

By comparison *linear* growth, as expressed by the catalytic path, would be the equivalent of placing a *single* grain of rice on each of the 64 squares. Hence the total amount of rice placed on the chess board would be just 64 grains! That's 64 grains of rice (representing catalysis) compared to some 2×10^{19} grains (representing autocatalysis). Autocatalysis is clearly an extraordinary reaction, explosive in its impact.

But do autocatalytic reactions actually exist? The answer is yes, they do, and in fact they are quite common in chemistry. For example, the reaction of acetone with bromine to give bromoacetone and hydrogen bromide is autocatalytic. That is because the reaction is catalysed by the presence of acid, and one of the products (hydrogen bromide) is an acid. Not surprisingly, the rates at which autocatalytic reactions proceed increase dramatically as the reaction progresses. However, that kind of autocatalytic reaction is not of special interest to us here. It is another kind of autocatalytic reaction, first discovered some forty years ago that is truly remarkable and enormously significant. I am referring to long chain-like molecules that are capable of making copies of themselves, molecules that are self-replicating. Sounds miraculous? It isn't—it's just chemistry. In 1967, Sol Spiegelman a microbiologist at the University of Illinois, performed one of the truly great classic experiments in molecular biology when he carried out molecular replication in a test tube.[27]

Spiegelman simply mixed an RNA strand (RNA stands for ribonucleic acid and differs slightly in structure from its more famous cousin, DNA) with free floating building blocks from which the RNA is itself built up, an enzyme catalyst to speed up the reaction, and lo and behold, the RNA strand ended up making copies of itself. Let us examine this replication reaction in greater detail. Self-replicating molecules, such as RNA, are self-replicating because they are able to induce a supply of building blocks, from which the molecule itself is composed, to connect up, thereby making a copy of the original molecule. A schematic representation of the RNA molecule is shown in Fig. 3a and the replication process is shown in Figs. 3b and 3c. From Fig. 3a we can see that RNA is a long chain-like molecule composed of segments called nucleotides that are linked together to make up that chain. In the case of an RNA molecule there are four possible nucleotides from which the chain may be built up, which can be simply labelled as U, A, G, and C. So an RNA chain might be represented by the sequence of those four letters, e.g., UCUUGAGCC... as indicated in the figure. Accordingly, the number of possible RNA chains, each with its particular sequence of nucleotides, grows dramatically as the chain length increases. Even for a relatively short RNA chain, say 100 nucleotides in length, the potential number of different chains is staggeringly large, 4^{100}. That's equal to 1.6×10^{60}—a 1 followed by 60 zeroes.

So how does a replicating RNA molecule manage to make an exact copy of itself from a mix of the four nucleotide building blocks and in just the right sequence, when the number of possible sequences is so staggeringly large? The answer lies in the ability of the RNA molecule to act as a template. What happens is that freely floating building blocks from which the RNA chain is composed, A, U, G, and C, latch onto the RNA chain as illustrated in Fig. 3b. Importantly, a lock and key type fit ensures that only the appropriate building block connects to any particular location on the RNA template so that the nucleotide sequence in the newly forming RNA chain is not arbitrary, but is specified by the original RNA strand; a U nucleotide latches onto an A segment in the RNA chain, an A nucleotide onto a U segment, a C nucleotide onto a G segment,

STABILITY AND INSTABILITY

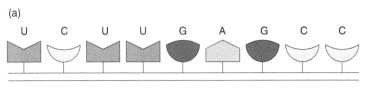

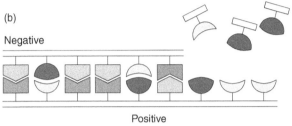

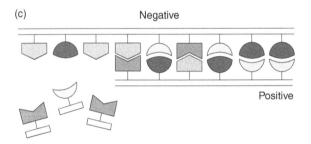

Fig. 3. (a) Schematic representation of an RNA molecule made up from a sequence of nucleotide building blocks, A, U, G, C. (b) Representation of the process by which an RNA chain induces a complementary copy of itself to be formed (positive to negative). (c) Representation of the process in which the complementary RNA copy induces a copy of the original RNA to be formed (negative to positive).

and a G nucleotide onto a C segment. Once the individual building blocks are all locked into place on the RNA chain, their proximity to one another enables them to link up so that a *dimeric* RNA entity results—*two* RNA strands weakly held together by bonds called hydrogen bonds. Because the bonds holding those two strands together are relatively weak, the two individual RNA strands can then separate, and two molecules of RNA now exist where initially there was only one. Of course these two strands are not identical, but complementary. Because of the lock and key interaction that binds the two strands together, U to A, G to C, the new strand can be thought of as a *negative* of the original strand, much like a photographic negative. But that means that once the negative strand acts to make a copy of *itself* in a *second* replication cycle, the resultant copy (a negative of a negative) is now a *positive*. So it is only after *two* cycles of template replication that the original RNA strand has in fact self-replicated, as indicated in Fig. 3c. So molecular self-replication reaction is a reality, a reaction that actually does take place, and, most importantly, is autocatalytic. It is autocatalytic because any self-replication reaction is by definition autocatalytic. And like the rice in the emperor–peasant story, the exponential growth that is often associated with replication reactions can result in the extreme amplification of even minute amounts of material, provided, of course, that the building blocks from which the replicating molecule is made up are available.

As a final point it should be noted that those individual building blocks, U, A, G, C, when mixed together in the *absence* of a template molecule do not readily link up into a chain. And even if they did, they certainly would not link up in one particular sequence. It is only when the RNA molecule acting as a template is added to the mixture that the nucleotide building blocks line up along the RNA chain in the proper sequence, lock into position, and link up, thereby causing a replica of the RNA chain to be created.

STABILITY AND INSTABILITY

Within living cells, molecular replication of the kind just described is actually quite routine. At the heart of every cell is the DNA molecule, that long chain-like entity in which the living creature's genes are located. A key component of cell division is the process of DNA replication so that each of the daughter cells, after division, has its own copy of the cell's DNA. In other words a single DNA molecule (barring copying errors) becomes two identical DNA molecules. But within a living cell that process of replication is a complex one as it takes place in a highly regulated manner and within a highly organized environment. Until quite recently molecular replication in isolation, without all the cellular paraphernalia to facilitate it, was unknown. Chemistry in all its variety and splendour did not include a category of self-replicating molecules, but in recent years that picture has changed dramatically. In fact in 1986, a dramatic step forward was taken when the leading German chemist, Günter von Kiedrowski, was able to carry out the first molecular replication reaction *without* any enzyme present to facilitate the reaction (i.e., no biological assistance)—finally pure replicative chemistry![28] Recall that Spiegelman's earlier replication experiment of the 1960s, though enormously significant, required the use of an enzyme to help the reaction proceed, and so was not purely chemical.

Let us then summarize the main chemical points so far.

1. Chemical reactions will only proceed if they are downhill in a thermodynamic sense such that less stable reactants are converted into more stable products.

2. Reactions that *are* allowed thermodynamically may not proceed, or may proceed slowly for kinetic reasons. An energy barrier has to be overcome for the reaction to take place.

3. Molecular self-replication of template-like molecules is an established chemical reaction and is kinetically unique. Being autocatalytic, self-replication can lead to dramatic exponential amplification of that template-like molecule until resources (building blocks from which the chain is composed) are exhausted.

The discovery that self-replicating molecules exist is highly significant because, as we will see, the existence of such molecules can form the basis for understanding how life emerged, how inanimate matter began the long and arduous road from simple beginnings to the extraordinary complexity that is life. Of course that single replicating molecule, whether RNA or some other related structure, does not in itself constitute life, not even simplest life. It is, after all, just a molecule. In fact, in many respects the reaction of self-replication is a chemical reaction governed by the rules of chemical reactivity, just like any other reaction. But there is something special about this self-replication reaction that leads us to believe it was the likely starting point for life. I have already indicated that self-replication, being autocatalytic, is kinetically unique in that it can lead to dramatic amplification, just like the effect of doubling the number of grains of rice on a chess board. We will now see how that kinetic power can lead us in quite unexpected chemical directions, in fact, to the establishment of a totally separate and distinct branch of chemistry, so distinct in its character that it goes under a separate label—biology! But in order to do so we first need to delve a little deeper into a basic concept of nature, one we have briefly mentioned in the context of the Second Law of Thermodynamics—the concept of stability.

STABILITY AND INSTABILITY

The nature of chemical stability

The concept of stability is a relatively straightforward and unambiguous one: an entity is stable if it persists, if it maintains itself without change over time. But here's the remarkable thing—within the material world stability can be of two fundamental and very different kinds—*static* and *dynamic*, one very obvious, the other rather less so. Static stability is the more obvious kind. For example, water, being a stable material in a thermodynamic sense, if suitably isolated, will remain unchanged over time, even over extended periods of time. Thermodynamic stability, which we discussed earlier, exemplifies this static kind of stability.

But there is another kind of stability—a *dynamic* kind, which is quite different to the static kind. Think of a major river, say the River Thames passing through central London. Its history can actually be traced back some 30 million years when it was a tributary of the River Rhine, but its current path and appearance are thought to have remained relatively unchanged for several thousand years. Accordingly, the River Thames, as an entity, may also be classified as quite stable. But in this case the kind of stability involved is very different from systems that are statically stable. The water that defines the River Thames *is not the same water, but is changing all the time*. The river we see today is in a sense a totally different river from the one we saw last time we looked. Its stability is a *dynamic stability*—the water that defines the river as a recognizable entity is constantly changing. A water fountain or a waterfall also manifests this dynamic kind of stability—the fountain (or waterfall) is stable (as long as the supply of water remains uninterrupted) but the water comprising that fountain (or waterfall) is being turned over continually.

So what does the stability of rivers, waterfalls, fountains, and the like, all displaying stability of a dynamic kind, have to do with chemical reactions, at least some of them? The answer is, quite a bit. Let's return to the reality of molecular replication. The process of molecular replication, because it can exhibit exponential growth, is *unsustainable*, just like doubling the grains of rice on a chess board. If one single molecule were to replicate 160 times it would (only in principle, of course) devour resources equal to the entire mass of the earth! What that must mean is that any replicating system (whether composed of replicating molecules, rabbits, or some other group of replicators) that *is* stable, can only be stable if its rate of formation is balanced (more or less) by its corresponding rate of decay. In other words, in order for the replication reaction to be maintained for any extended period, the replicating system has to decay at a rate that is commensurate with its rate of formation. Under those circumstances the replication process, in principle at least, can proceed indefinitely.

But what would cause replicating entities of whatever kind to decay? If the replicator is chemical, say a replicating molecule, then that molecule will undergo competing chemical reactions, so such molecules will not survive for too long. RNA oligomers (an oligomer is just a chain-like molecule made up of component building blocks) and peptides, the prime examples of molecules capable of replication, are not too stable thermodynamically speaking and constantly undergo degradation processes. And if the replicating entity is biological—a bacterium or some multicell creature, the situation is much the same. In this case decay (now termed death) is also lurking close by. Lack of nutrition, chemical or biological attack, physical damage, apoptosis (programmed cell death), or other mechanisms, will eventually lead to the demise of all living things. The eventual death/decay of all living things, by whatever mechanism, will therefore balance the ongoing replicator formation and facilitate the dynamic stability of the replicating system.

The important point, however, is that if a replicating system is found to be stable over time, it is the *population* of replicators that is stable, not the *individual* replicators that make up that population. The individual replicators are being constantly turned over just like the water droplets that make up the river or fountain. In other words, the stability associated with a stable population of replicating entities, whether molecules, cells, or rabbits, is of a *dynamic kind*, just like that of the river or fountain. Think therefore of a stable population of replicating molecules as a *molecular fountain*. We will see how life's dynamic character, a feature that has troubled modern-day biologists, derives directly from the dynamic character of the replication reaction.

In the context of chemical systems, static and dynamic forms of stability are very different. In the 'regular' chemical world a system is stable *if it does not react*. That is the very essence of stability—lack of reactivity. In the world of replicating systems, however, a system is stable (in the sense of being persistent and maintaining a presence) *if it does react—to make more of itself*, and those replicating entities that are *more* reactive, in that they are better at making more of themselves, are *more* stable (in the sense of being persistent) than those that aren't. This is almost a paradox—greater stability is associated with greater reactivity. We therefore call the kind of stability associated with replicating systems a *dynamic kinetic stability*. Its stability is dynamic for the reasons we have outlined, but we need to introduce an extra term in the description—the word 'kinetic'—to distinguish it from the dynamic stability of fountains, rivers, and the like, which is physical, and not chemical. For replicating systems the *rate* at which the replicating system makes more of itself, together with the rate at which it decays, are key parameters in determining the level of stability. High stability will be facilitated by a *fast* rate of replication and a *slow* rate of decay since that will lead to a large population of replicators. To our chagrin, mosquitoes and cockroaches are highly stable in this dynamic kinetic sense—they are extremely efficient in maintaining a large population, whereas pandas, for example, are much less efficient. Indeed, low dynamic kinetic stability for a replicating entity, whether due to slow replication or fast decay, may well lead at some point to the population of that replicator becoming extinct.

I have described here the existence of a distinct kind of stability quite different from the regular stability with which we are more familiar, so given the existence of two kinds of stability, one might ask which is the preferred one, which stability is inherently the more 'stable'? A definitive answer to the question is actually not possible—it's the old apples and oranges problem. The two kinds of stabilities are not directly comparable and in fact one of them, dynamic kinetic stability, is only quantifiable in a very limited way. But intuitively we might suspect that static stability, the one based on a lack of reactivity, is inherently the preferred kind of stability, the one likely to be more enduring—wouldn't it? Well, not necessarily! In examining the world around us we are led to a surprising conclusion. Mt. Everest, for example, a statically stable entity (ignoring tectonic movements), is thought by geologists to have existed for some 60 million years, so clearly static stability can be very substantial. But cyanobacteria (blue-green algae), a very ancient life form, appear to have continuously populated the earth for several billion years, with little, if any morphological change. Biologists might argue over the period that they have remained unchanged, whether it is closer to 2.5 or 3.5 billion years, but there is no argument that cyanobacteria have been around for several billion years. Now that *is* stable! Of course, we are speaking here of a dynamically stable system—the cyanobacteria alive today are not the same ones that were alive several billion years ago. But through ongoing replication they have maintained a continual presence on this planet for an extraordinarily long period of time. Let us be clear: despite the dynamic character associated with replicating systems, their form of stability should not be underestimated; it is able to encompass time frames that cover a significant fraction of this planet's 4.6 billion-year lifetime.

Our discussion till now has made clear that (static) thermodynamic stability and dynamic kinetic stability are applicable to different systems and are quite distinct in their nature. But the fact that there are two very different kinds of chemical stabilities has profound implications for both the physical and chemical characteristics of systems within each of the two classes. This is because the rules governing transformations for chemical systems belonging to the two different stability types are necessarily different. In effect there are *two* chemistries out there! One of the chemistries is just 'regular' or traditional chemistry, which has been studied for several centuries and is well understood—a mature science. The other is replicative chemistry, the chemistry of replicating systems. This other chemistry, part of a new area of chemistry recently named 'systems chemistry', is still in its infancy.[29] Systematic study in the area was only initiated some twenty-five years ago and many chemists remain unaware that such a field even exists. Let us now flesh out the nature of this 'other chemistry', why it comes about, what are some of its prime characteristics, and how this new field is providing the basis for the building of bridges between the sciences of chemistry and biology.

Rules governing replicator transformation

In 1989, Richard Dawkins alluded to a fundamental law of nature which applies to both the biological as well as the broader physicochemical world: the *survival of the most stable*.[30] Steve Grand, in his 2001 book, *Creation*, expressed it somewhat differently: *Things that persist, persist. Things that don't, don't.*[31] This sounds like a tautology, and in some respects it is. But there is an important message present within that seemingly trite statement. Once it is (empirically) evident that matter is not immutable, that it is susceptible to chemical change, then it necessarily follows that matter will tend to be transformed from less persistent to more persistent forms, in other words, *from less stable to more stable*. Persistent forms don't tend to change because they are...persistent. And, of course, *less* persistent things *do* tend to change because they are *less* persistent. So matter, by definition one could say, tends to become transformed from less persistent to more persistent forms, or couched in stability terms, from less stable to more stable forms. As a matter of fact, that is what chemical kinetics and thermodynamics is all about—being able to explain or, better still, predict the likely reactions of chemical systems in their search for more stable forms. And what is the central law that governs such transformations? The Second Law. A mixture of hydrogen and oxygen gases readily reacts

STABILITY AND INSTABILITY

to give water because the hydrogen-oxygen gas mixture is unstable, whereas the water product is stable. When matter reacts chemically, it reacts so as to become transformed from thermodynamically *less* stable reactants into thermodynamically *more* stable products.

But what happens in the chemical world of replicators, in the world of replicating molecules, for example? What rule governs transformations within that world? Of course a replicating molecule may undergo chemical reactions in which it is converted into one or more *non-replicating* molecules. However, we are not concerned here with those kinds of reactions. They are covered by the rules governing chemical reactions generally. The reactions that are of special interest are those in which a replicating molecule (or set of molecules) is transformed into some *other* replicating molecule (or set of molecules). It is these reactions, which address the nature of replicating systems *as a class*, that we must further explore. As we will discover, it is this very special class of molecules that offers unique potentialities. And now to the essential point: given that the kind of stability applicable in the replicating world is dynamic kinetic, not thermodynamic, the rule that *effectively* controls transformations within the world of replicators is *not* the Second Law, but one that is expressed in terms of dynamic kinetic stability. The rule is simply stated as follows:

> *Replicating chemical systems will tend to be transformed from (dynamically) kinetically less stable to (dynamically) kinetically more stable.*

That selection rule is in some sense an analogue of the Second Law, the selection rule in the regular chemical world. In both worlds chemical systems tend to become transformed into more stable ones, but as the two worlds are each governed by a different kind of stability, the selection rule in each world is different—thermodynamic stability in the 'regular' chemical world, dynamic kinetic stability in the replicator world. As we will now see, the implications of those distinct selection rules are profound. But before we discuss those implications, is there any evidence for the distinctly different selection rule in the replicator world that is being proposed here? Yes, there is. Back to Sol Spiegelman and his remarkable RNA replication experiment conducted over forty years ago.

In describing Spiegelman's landmark experiment earlier in this chapter, I neglected to tell the whole story. It is true that an RNA strand when mixed with its component building blocks (and added enzyme catalyst) undergoes a self-replication reaction. But something else takes place as well, something of considerable significance. Replication may on occasion occur imperfectly, in that the wrong nucleotide segment will attach to the template. For example, a C nucleotide, rather than an A nucleotide, will attach to a U segment on the template chain. Thus, on occasion, *imperfect* replication will lead to the formation of a *mutant* RNA strand. In other words, over time the solution will begin to consist of both *original* RNA strands as well as *mutated* ones. And here Spiegelman made a remarkable observation. Over time the solution began to be populated by mutant RNAs that replicated *more* rapidly than the original RNA strand. In fact the original sequence after some time may even disappear from solution! In other words, a process akin to Darwinian selection was found to take place at the molecular level—the RNA strands evolved. Since short RNA strands replicate more rapidly than longer RNA strands, the initial strand composed of some 4,000 nucleotides began to shorten and eventually ended up with just some 550 nucleotides. The replicating prowess of the short strand was so dramatic it was termed Spiegelman's Monster!

Before continuing it is important to note that the evolutionary process observed by Spiegelman is chemical in essence, not biological. An RNA strand in no way constitutes a living entity—it is a molecule; admittedly a biomolecule, meaning that it is a molecule of the kind normally found in living systems, but a molecule is a molecule is a molecule. And the fact that a slowly replicating molecule tends to evolve into a more rapidly replicating one is due to chemical factors, chemical kinetics to be precise. Nothing biological here—just chemistry. While this is not the place to go into a detailed kinetic analysis of the competition between two replicating molecules, the bottom line is easy to state. When a number of different replicating molecules all compete for the common building blocks from which they are constructed, the faster replicators out-replicate the slower ones so that over time the slower replicators will tend to disappear. What effectively happens is that slower replicators are replaced by faster ones in precise agreement with the general selection rule for replicating entities that was specified above.

As a final point it is important to ask how the two stability kinds, static and dynamic kinetic, interrelate. The statement that the replicating world is governed by the drive toward greater dynamic kinetic stability, though correct, needs to be qualified, and that qualification can be expressed through the metaphor of Russian dolls. Although the replicative world is governed by an analogue of the Second Law, no physical or chemical system can avoid complying with the Second Law itself. That is the grand and comprehensive rule, the one governing *all* transformations in the material world. So how can two different laws operate simultaneously on the one system? The answer is that the Second Law analogue governs replicating systems within the constraints of the Second Law itself, just like Russian dolls that fit one within the other. A simple example from everyday life may clarify the issue.

Your car breaks down and you ask your mechanic to explain the reason for that breakdown. If he mumbles something about the Second Law of Thermodynamics as the explanation for the breakdown, you'd be quite frustrated, even though his explanation was entirely correct. Correct, but quite unhelpful. The direction of all irreversible processes is governed by the Second Law, so whatever event caused your car to break down it was in a fundamental way governed by the operation of the Second Law. So why was the answer unsatisfactory? Because there are rules that govern car function—how engines operate—that sit within the more general framework of material happenings as expressed by thermodynamics. The Russian doll of engine function sits within the bigger thermodynamic doll. To fix your car you want to know what happened within the context of the smaller doll, the one that deals specifically with engine function. Did the fuel line become blocked or did the timing belt break? The Second Law, the more global explanation, though correct, is of no practical use. And so it is with replicating systems. Stable replicating systems operate according to the rules that govern replicating systems, as described earlier in this chapter, but that specific behaviour is not *independent* of the Second Law. Rather, it operates *within* the general constraints that the Second Law places on *all* material systems. There is no contradiction then between the two rules. The underlying message in the Russian doll metaphor is that we will be better able to understand reactions in the replicative world by considering the rules that govern *that* world, rather than the more general thermodynamic principles that govern *all* material systems. Stating that the reactions of replicating molecules and biological evolution, in general, are governed by the Second Law is formally correct, but very much like saying that that is the reason your car broke down. Correct, but not particularly helpful!

Though this chapter on chemical stability and reactivity was chemical in its approach, we will subsequently see that it can provide the basis for our attempt to bridge between chemistry and biology. We will discover that biological terms, such as fitness, are directly related to chemical terms such as stability. But before we seek to understand the chemistry–biology connection in depth and to discover the relationship between chemical replicators and biological ones, let us consider the process which necessarily led to the transformation of chemistry into biology—the origin of life on earth—and see why that issue continues to remain stubbornly controversial. As I have already pointed out, if we want to understand what life is, we have to understand the essence, if not the detail, of the process by which it came about.

5

The Knotty Origin of Life Problem

Mankind's preoccupation with life and its origin can be traced back almost 3,000 years and it is not by chance that the opening lines of the first book of the Old Testament, Genesis, offers a biblical account of that extraordinary event. The narrative from there is long and convoluted, but we will take up the story from the early twentieth century, which is when the modern scientific dialogue commenced in earnest.

The origin of life problem is a tantalizingly difficult one. George Whitesides, the distinguished Harvard chemist recently expressed it in unusually frank terms: 'Most chemists believe, as do I, that life emerged spontaneously from mixtures of molecules in the prebiotic Earth. How? I have no idea.' That sums it up pretty well. In this chapter I will review this long-standing question to see where the debate currently stands and where the major problems lie. My approach is critical rather than comprehensive.[32]

Let us begin with a fundamental but unproven assumption, that life on earth was initiated from abiotic beginnings some period of time after our solar system was formed some 4.6 billion years ago. That assumption forms the basis of the modern view which took shape in the 1920s through the joint contributions of the Russian biochemist, Alexander Oparin, and the influential British geneticist and evolutionary biologist, J. B. S. Haldane. An alternative scientific view, panspermia, invokes the idea that life originated from beyond the earth and was transported in some fashion to the prebiotic earth. That idea, proposed toward the beginning of the twentieth century by a well-known physical chemist, Svante Arrhenius, is, however, no longer seriously considered by the majority of researchers in the area, even though it has been supported by some well-known figures, including Francis Crick. A key difficulty with the panspermia proposal is that it does not really solve the problem of abiogenesis—the manner by which life emerged from inanimate beginnings—it merely transplants the problem to some other unidentified cosmic location. Regardless of its location, the question remains unchanged: how did life emerge from non-life?

Historical and ahistorical approaches

Before addressing the origin of life question in greater detail, it is crucial to point out that the question has two quite distinct facets—*historical* and *ahistorical*, and only the *combined* insights of the two facets will be able to lead to a full and satisfying resolution of the problem. The historical aspect would seek to answer the *how* question—*how* did life emerge. That would involve deciphering the actual chemical events that transpired on the prebiotic earth—the particular chemical path followed, step by step, leading from inanimate materials through to simplest life. Key questions would include: what were the molecular building blocks from which early life was constructed? What were the prevailing reaction conditions that enabled those building blocks to form, and once formed, what were the key intermediate steps along the long evolutionary road from those building blocks to simple life? As we will shortly see, not only has no broad agreement on these issues been reached, but practical knowledge of any kind regarding specific conditions on the prebiotic earth remains seriously wanting.

The ahistorical aspect would address the more general *why* question: why would inanimate matter of any kind, regardless of its structural identity, follow a pathway of complexification in the biological direction, eventually leading to some simple life form? I ask the question in the sense of identifying a driving force, seeking the same kind of insight that Newton sought when he asked 'why do apples fall?' Could the process, at least in principle, be induced in a range of different materials? What physicochemical principles could explain such an extraordinary chemical transformation? And with respect to this last question, can we go a step further and postulate the existence of a 'physical' driving force that would have directed inanimate matter to complexify in the biological direction? That question, as phrased, rests on an additional presumption, that the emergence of life was not a purely random event, but one that was induced by established physicochemical forces. We will discuss this presumption in greater detail subsequently. Thus the ahistorical perspective would not focus on the precise molecular identities of the relevant inanimate materials, but would seek out generalities—the *category* of material (or materials) that would likely have the propensity to become life, as well as the relevant physicochemical principles which would have induced

THE KNOTTY ORIGIN OF LIFE PROBLEM

these materials to complexify into a simple life form. As we will see, here too the picture remains uncertain and highly controversial.

Given the paucity of information of any kind on the origin of life, ideally we would want to optimize our insights from both historical and ahistorical aspects in order to obtain as full a picture as possible. Before doing so, however, it will be useful to clarify the manner in which the two aspects interrelate. It turns out that each kind of information can serve as a means of obtaining information about the other kind. To illustrate how this interrelation works let us consider a simple physical analogy—a boulder that rolls down some mountain slope after being dislodged from its initial location (let us say due to water erosion or seismic activity), and finally comes to rest at the bottom of the slope. In this case ahistorical and historical aspects of that physical event are readily identified. The historical *how* question would be: from what initial location did the boulder begin its descent, and what trajectory did it follow? In principle there could be a large number of possible starting points, with each matched to a large number of potential trajectories. The second question—the ahistorical *why* question—would be to ascertain why, once dislodged, the boulder was induced to roll down the slope in the first place. Of course, in the case of a rolling boulder the answer to the second question is obvious—a gravitational force operates on all objects on the earth's surface tending to lower their potential energy, so the physical reason for the boulder rolling down the slope is clear. Notice that the answer to the 'why' question is formulated in terms of a general law, independent of the specific location of the boulder, the nature of the terrain, etc.

There is an important message hidden within the rather trivial rolling boulder analogy. Historical knowledge and ahistorical understanding impact on one another. For example, understanding the ahistorical aspect—the nature of the physical force responsible for boulder motion—would greatly assist in answering the historical question, that of boulder trajectory. Indeed, due to our knowledge of gravity we can safely exclude the possibility that the boulder simply levitated and floated through the air to its final location. Only boulder trajectories consistent with the action of a gravitational force would merit consideration. But the historical-ahistorical interplay also operates in reverse. Let's assume for a moment that we are *not* familiar with the law of gravity. Obtaining information regarding the boulder's trajectory would provide information as to the principles governing boulder motion in general. Uncovering the boulder's trajectory would reveal that boulders apparently seek to lower their potential energy, i.e., they always roll from a higher location to a lower one—never in reverse, never uphill, and that the preferred trajectory is the pathway of steepest descent. So knowledge of a particular boulder's trajectory, a historical event, would be a key step toward uncovering the rules governing boulder motion in general, the ahistorical aspect.

In the same way, if we want to address the historical origin of life question in its particulars, namely, to specify the starting materials and the particular set of reaction steps that led from those materials to early life forms, then knowing the general principles that govern the conversion of relatively simple molecular systems into the complex systems of life would be of considerable value. It would suggest the kinds of historical evidence we should be seeking. And vice versa, knowing the reactions that led to the conversion of inanimate matter into animate matter would greatly assist in uncovering the general principles that would have governed such a remarkable transformation.

THE KNOTTY ORIGIN OF LIFE PROBLEM

But it is precisely at this point that we run into difficulty. Both kinds of information with respect to the origin of life problem are seriously lacking. Let me spell it out: since we are still struggling to understand the ahistorical principles, we don't know which historical processes we should be seeking to uncover, and since we have no definitive historical evidence for the emergence of life at a particular location under particular conditions, we have no historical data to guide us in the elucidation and formulation of the relevant ahistorical principles. Catch 22!

So how to proceed? Before doing so, I will make a somewhat controversial statement with respect to the origin of life problem, *that the ahistorical question is the more significant one scientifically, and also the inherently more tractable one, the one that is less difficult to resolve.* As I will explain subsequently, this presumption will impact considerably on the nature of our discussion. I will describe how attempts to trace out plausible historical mechanisms for the emergence of life, without a prior understanding of the principles governing biological complexification, have not been able to resolve the problem, and may have even contributed to existing confusion. Often these hypotheses are untestable and, being highly specific in their formulation, do not address the more general ahistorical question. With this brief introduction, let us now examine the topic in some detail.

History of life on earth

What historical information do we have regarding early life on the planet? On the basis of radiometric dating it is generally accepted that the earth was formed some 4.6 billion years ago, with the first 600–800 million years of the planet's existence being thought to have been too inhospitable for the emergence of life. During that initial period extensive bombardment from outer space would have been capable of evaporating the oceans and sterilizing the earth's surface. The earliest historical evidence for the existence of life on earth is obtained from what is termed the palaeobiologic record—the microfossil remains of ancient microorganisms, and most recent findings date that early life at about 3.4 billion years old.[33] In any case all of these fossil findings point to relatively advanced cellular life, and therefore do not throw light on the earlier process of abiogenesis. Indirect evidence for the existence of earlier life going back 3.8 billion years ago is also claimed[34] though the issue remains controversial. In other words, the morphologically informative palaeobiologic record runs out after about 3.4 billion years when cellular life was already well established, so we must conclude that the palaeobiologic record in itself is unable to provide direct insights into the origin of life problem.

The second powerful technique for probing the history of life on earth is termed phylogenetic analysis, or, by its simpler name, sequence analysis. As the Nobel biophysicist, Max Delbrück, noted in a 1949 address before the Connecticut Academy of Arts and Sciences: 'any living cell carries with it the experience of a billion years of experimentation'. In that spirit, sequence analysis enables us to construct a *tree of life*, which reveals the way all living things relate to one another by uncovering their evolutionary history. At the base of the tree we place the Last Universal Common Ancestor (LUCA)—the most recent living thing from which *all* life on earth proceeded to evolve—and from that base the trunk divides continually into more and more branches, each branch representing a new species. The very top of the tree represents all living species found on the earth today, while lower branches that end abruptly represent species that became extinct. So start from any branch end, follow it back, and you will uncover the entire evolutionary record of that particular species.

How does sequence analysis reveal the structure of the tree of life? Two families of biomolecules that govern the nature and form of all living entities are nucleic acids and proteins, already mentioned in earlier chapters. Both of these groups of compounds involve long, chain-like molecules made up of monomeric units—nucleotides in the case of the nucleic acids, and amino acids in the case of the proteins. Since there are a variety of possible monomeric units from which the nucleic acid and protein molecules can be built up (4 nucleotide possibilities in the case of nucleic acids, 20 amino acid possibilities in the case of proteins), considerable variation in the *sequence* of the monomeric units within these two classes of biomolecules is possible. But here is the important point: the closer two species happen to be in evolutionary terms, the greater the similarity in the sequence of any biomolecule shared by the two species is likely to be. Carry out that comparative study for a large number of species and the genealogical relationship between different life forms, as expressed in a tree structure, can be established.

Sequence analysis in the 1970s started off with a spectacular result. Before those studies it had been assumed that *archaea* and *bacteria*, two single-cell life forms, were closely related based on their similar morphologies and prokaryotic nature (primarily, the absence of a nucleus and other organelles). Archaea are commonly found in relatively harsh environments, such as hot springs and salt lakes, where traditional and more common bacterial life forms cannot survive. But in the late 1970s, primarily due to the pioneering sequence analysis work of Carl Woese, it was discovered that archaea are more closely related to eukaryotic cells (those making up you and me) than to bacterial cells! As a result, the two seemingly closely related prokaryotic life forms were relegated to distinct and separate kingdoms. The tree of life, thought to be composed of *two* major kingdoms, Prokarya and Eukarya, was transformed into a tree with *three* kingdoms—Archaea, Bacteria, and Eukarya (illustrated in Fig. 4). Sequence analysis had proven to be a most powerful tool in elucidating genealogical relationship. A major step in constructing the tree of life had been taken. But that's where the good news stops. Applying that powerful tool to the origin of life problem has proved disappointing. Sequence analysis has failed to throw more than minimal light on that problem. Let's see why.

In recent years the true significance of sequence analysis, even for established life forms, has been increasingly questioned. The problem initially arose in the 1990s when it became possible to carry out complete genomic (DNA) sequencing and not just sequencing based

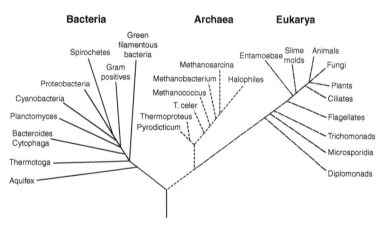

Fig. 4. Tree of life based on ribosomal RNA sequence analysis showing three kingdoms of life—Bacteria, Archaea, and Eukarya.

on RNA and protein. The troublesome finding was that the tree topology using the different molecular probes often differed significantly. One tree might suggest, for example, that species A is closely related to species B but not to C, whereas the other method might suggest that A is more closely related to C, and not to B. Clearly conflicting topologies cannot all be right. The primary explanation for this anomaly was quickly understood to be Horizontal Gene Transfer (HGT),[35] the process in which an organism transfers genetic material to some organism other than one of its own offspring. HGT contrasts with traditional vertical gene transfer, whereby gene transfer takes place in the traditional fashion—from parent to offspring, the way heredity normally operates. The result is that the genealogical significance of the particular tree topology that is obtained cannot be entirely assured—the tree outline starts to blur.

For established life forms the extent of the problem is a subject of ongoing debate, but with regard to the origin of life issue the news is worse. The problem is that the further back one follows the branches of the tree, the greater the impact of HGT seems to be. In fact, Carl Woese, whose life work focused on such phylogenetic analysis, argues that early cellular organization would have been loosely connected and modular, that evolution would have been communal, not individual, so that such entities would not have even had stable genealogical records.[36] This then suggests that the root of the universal tree of life, LUCA, may be something of an artefact which derives from forcing a tree representation on the sequencing data. If true, that statement has important consequences. It means that the nature, or even the very existence, of a discrete LUCA remains uncertain. That of course makes any phylogenetic extrapolation regarding pre-LUCA entities even more questionable. After almost four decades of phylogenetic analyses the methodology has had to undergo significant reassessment. The tree of life, at least for archaea and bacteria, has been replaced by a web of life. A tree topology of course leads back to a trunk and to roots, but a web topology, unfortunately, does not lead anywhere; a web topology is not a useful source of historical information. The bottom line: when the significance of phylogenetic analyses of established life forms and the LUCA, in particular, are increasingly being questioned and revised, extracting useful phylogenetic information regarding *earlier* transitional life forms (pre-LUCA) seems to be, at least at present, a questionable endeavour.

We have discussed the palaeobiologic and phylogenetic tools as a means of obtaining historical information of early life on earth and found that they are unable to provide insights into the process by which inanimate matter was transformed into simple life. However, there is an additional approach to the historical question that potentially could provide useful information: assessing the kind of prebiotic chemistry that could have taken place on the earth, given prevailing prebiotic conditions. Could the study of prebiotic chemistry provide insights into life's beginnings? Regrettably, the answer to that question has also not been encouraging. Despite considerable effort that has gone into exploring this line of thinking, the fruits of that labour have been meagre. Let us now look at the main contributions to that effort and consider why they have met with limited success.

Prebiotic chemistry

It is clear that for life to have emerged on earth, the appropriate building blocks, from which all living systems are constituted, must have been available. Accordingly, it seems reasonable to presume that some hints with regard to the origin of life could be revealed through analysis of the materials that might have been formed on the prebiotic earth. Though a 1924 paper entitled 'The Origin of Life' by Alexander Oparin offered some early ideas on the prebiotic formation of organic materials, the origin of life question was thrust into prominence with the landmark experiments of the American chemist, Stanley Miller.[37] In these experiments Miller, then a graduate student under the direction of Harold Urey, a Nobel chemist at the University of Chicago, took a mixture of the four gaseous components thought at the time to be the main constituents of the prebiotic atmosphere—hydrogen, ammonia, methane, and water vapour, and simulated the effect of primordial lightning by passing an electrical discharge through the mixture.

The result was dramatic. A range of organic materials, including a number of amino acids, were found to have been formed. Since amino acids are the building blocks of proteins, proteins being a key component of all living systems, a new area of study was established—the field of prebiotic chemistry, a field that quickly became a focus of considerable scientific interest. The prevailing thinking was that by conducting additional Miller-type experiments under presumed prebiotic conditions, the source of other key life components might be uncovered, thereby contributing to the resolution of the origin of life problem. Indeed, within a few years another group of organic substances, the organic bases, which constitute a key component of all nucleic acids, were also shown to be readily synthesized from available simpler materials, under what were considered to be likely prebiotic conditions. For a period the road ostensibly leading to the origin of life began to look like a superhighway.

But not for long. Dissenting voices quickly arose. Just where on the earth did life's emergence take place? The initially preferred location, within a so-called 'prebiotic soup', was questioned for a variety of reasons and the hunt for creative alternatives quickly expanded. Two of the more prominent ones were the suggestion that life originated in hydrothermal vents deep under the sea,[38] while another proposed that life was initiated on clay surfaces.[39] Differences don't get much greater than that! But then questions regarding the composition of the prebiotic atmosphere arose. Was the prebiotic atmosphere in fact reducing, as initially proposed, or, on the basis of more recent data, was it neutral, containing mainly carbon dioxide, nitrogen, and water? No broad agreement on any of these fundamental questions seems to have been reached.

Thus the initial excitement induced by Miller's experiments was gradually replaced by a phase in which a range of competing, mutually incompatible proposals were offered. Optimism gave way to lack of coherence and uncertainty. In fact, the only point on which the different mechanistic proposals for the emergence of life *were* in agreement was that life on earth *did* emerge some 4 billion years ago from inanimate materials present on the prebiotic earth. It is true that the richness of chemistry associated with prebiotic styled experiments did lead to the discovery of a range of novel chemical reactions and opened up alternative ways of thinking about the topic. However the considerable effort that was put into that endeavour seemed to have been accompanied by a questionable way of reasoning. In simplest terms, a general thesis that formed the basis for much of the discussion on prebiotic chemistry took shape, namely, that from the study of chemical reactions under supposed prebiotic conditions, it is possible to outline pathways that could have led to the emergence of life. In retrospect that thesis now appears to be highly problematic. Seeking out the historical conditions for life's beginnings on the prebiotic earth has not contributed significantly to resolving the origin of life problem.

There are several problems with the 'prebiotic chemistry' approach. First, the absence of reliable information regarding conditions on the prebiotic earth, certainly with respect to any specific location, has significant consequences. If we want to specify the nature of reactions that could, or could not, have occurred at some particular site on the prebiotic earth, the available materials and the corresponding reaction conditions at that site must be specified. But since neither the available materials nor the reaction conditions are known, almost nothing can be said with any degree of confidence.

To illustrate the depth of the problem, consider for example the expression 'conditions on today's earth', an expression presumably more definitive than the corresponding term 'conditions on the prebiotic earth'. But what does 'conditions on today's earth' actually signify? Are we speaking of the conditions within an erupting volcano, under the arctic ice shelf, at the bottom of the ocean, in a hydrothermal vent, in the hot sands of the Sahara desert, in a freshwater lagoon, or in any number of other totally different locations? The term raises considerable uncertainty even though we *can* specify with some precision the conditions at any given location. But when we speak of conditions on the prebiotic earth, and do so in a most general way, the uncertainty takes on an extra dimension. Not only don't we know *where* on earth particular prebiotic events took place, but we don't really know the actual conditions at any of those prebiotic locations. And to make things more difficult, the study of physical organic chemistry teaches us that reaction paths and reaction mechanisms can be quite sensitive to reaction conditions, so any proposals as to what may or may not have taken place at some point on the prebiotic earth can only be classified as highly speculative.

Speculation on these questions is also methodologically problematic since it is unlikely that any scenario is falsifiable in practice. The number of plausible scenarios would only be limited by the creative efforts of those chemists applying themselves to the question. Needless to say the lack of falsifiability necessarily undermines the utility and significance of any particular proposal. As Leslie Orgel, the eminent British chemist and leading origin of life researcher, once put it: 'Just wait a few years and conditions on the primitive Earth will change again.' A cynic might argue that here we have the ideal research area. One could safely publish in the field, secure in the knowledge that no one is ever likely to prove you wrong!

There is a second problem, no less fundamental, with the presumption of particular prebiotic conditions. Even if prebiotic conditions could be specified with some precision, it has been frequently assumed that the knowledge of such conditions would enable us to specify not only what reactions *could* have taken place, but also what reactions *could not* have taken place. That presumption has in fact been used to argue against one of the main origin of life scenarios—the existence of an RNA-world as a transitional period on the way to simplest cellular life. Since long chain RNA molecules are formed from their component building blocks—RNA nucleotides—the RNA-world scenario crucially depends on the appearance on the prebiotic earth of those nucleotides. The argument offered was essentially the following: if, despite several decades of effort, gifted chemists were unable to synthesize RNA nucleotides under presumed prebiotic conditions, then it can be safely concluded that such nucleotides could not have spontaneously appeared on the earth.

Here the flawed logic is easily exposed. We simply cannot rule out the possibility of prebiotic RNA nucleotides emerging spontaneously because, as the old saying goes: absence of evidence does not constitute evidence for absence. How many decades of effort by gifted chemists are required before the conclusion is justified? Two, three, maybe five? And how gifted do the chemists have to be? As discussed above in some detail, it is simply unreasonable to conclude that prebiotic conditions at *every* location on the early earth would have precluded the emergence of nucleotides when the available materials and reaction conditions at *any* of the possible locations remains unknown.

In any case, the fallacy was laid bare quite recently by the imaginative British chemist John Sutherland when he did the 'impossible'. John Sutherland *was* able to synthesize an RNA nucleotide from so-called prebiotic starting materials and the breakthrough came about by his thinking out of the box, by utilizing a novel synthetic strategy quite different from the conventional one attempted by earlier researchers.[40] One can only fantasize as to how many other feasible 'prebiotic syntheses' of nucleotides or any other key building block might in principle exist. Shouldn't nature also be allowed the prerogative of 'thinking out of the box'? The conclusion is clear: though one can safely conclude from experimental results which chemical reactions *are* possible, it is logically unsound to conclude what reactions are *not* possible, what *could not* have taken place, particularly over a time span of hundreds of millions of years, and under effectively unknown reaction conditions. The comment by a pioneer in the origin of life area, the venerable Peter Schuster, regarding prebiotic chemistry is particularly apt: 'Never say never!' We will return to the possible role of RNA on the prebiotic earth in chapter 8, as the fortuitous emergence of a molecule capable of self-replication is a central theme in the origin of life debate.

The above two arguments have demonstrated the inherent difficulties in the prebiotic chemistry approach to the origin of life. However it turns out that the problems run even deeper. We argued above that seeking to discover the prebiotic conditions that could have led to the emergence of biologically relevant materials is problematic. But behind that endeavour lies an unstated assumption, namely, that if some convincing explanation for the availability of the key biomolecules from which all living things are composed—sugars, bases, nucleotides, amino acids, lipids, etc.—can be found, then a major step toward resolving the origin of life problem will have been taken. Unfortunately that assumption is also questionable. Even if all the experiments in prebiotic chemistry had been carried out with total success, thereby fulfilling prebiotic chemists' wildest dreams, the origin of life riddle would still be a riddle, because the true problem with regard to the origin of life goes beyond the question of how life's building blocks appeared on the prebiotic earth. A deeper problem lies elsewhere.

Consider, you establish a group of leading biochemists, synthetic chemists, molecular biologists, and you ask them to create a simple living system in their laboratory. No restrictions of any kind, no chemical limitations, none of the constraints that would have necessarily accompanied conditions on the prebiotic earth. And no funding limitations either! Offer them whatever materials they would like in any combination they would like—DNA and RNA oligomers, lipids, assorted proteins, sugars, any catalyst they would want, and, of course, any instrumentation they might require. Create for them any reaction conditions needed to carry out their experiments, prebiotic or otherwise. If they request simulated conditions resembling those within a hydrothermal vent, no problem. Clay surfaces? That one's easy. But the honest response? Most would not really know where to start!

Certainly, a number of audacious scientists, such as Jack Szostak, the Nobel geneticist at the Harvard Medical School, and Pier Luigi Luisi, the venerable Italian chemist, have taken some tentative steps toward that ambitious goal,[41] but for reasons we will discuss in chapter 8, the obstacles in reaching the target remain formidable. The problem of how life emerged on the prebiotic earth is not just about what materials were available and identifying the reaction conditions at the time, because even the very best chemists without any resource limitations would not really be sure how to proceed. And the problem does not stem from the fact that one particular step or other in the recipe for life is especially difficult and still technically out of reach. The problem is more fundamental. The problem is there is still no coherent recipe. As we noted earlier, we don't yet adequately understand what life is, so how can one go about making something that we do not as yet fully understand? So, in a fundamental sense, the efforts to uncover prebiotic-type chemistry, while of considerable interest in their own right, were never likely, in themselves, to lead us to the ultimate goal—understanding how life on earth emerged.

In fact we would go so far as to say that seeking historical information regarding the emergence of life on earth is a honey trap—seductively appealing, beckoning both the novice and the experienced researcher, but one that is unlikely to yield genuine insights with respect to the question it poses. More significantly however, historical evidence alone, even if it were to become available, would not resolve the problem. The real challenge is to decipher the ahistorical principles behind the emergence of life, i.e., to understand why matter of any kind would tend to complexify in the biological direction. It is this ahistorical question, independent of time and place, which lies at the heart of the origin of life problem. In order to resolve the origin of life mystery, and it is a mystery, we need an understanding of the physicochemical processes that would have converted inanimate matter of whatever kind into a chemical system that we would categorize as living. *That* is the issue that kept the great twentieth-century physicists awake at night, not prevailing uncertainties with regard to the composition of the prebiotic atmosphere or the feasibility of synthesizing nucleotides under prebiotic conditions, and the like. What laws of physics and chemistry could explain the emergence of highly complex, dynamic, teleonomic, and far-from-equilibrium chemical systems that we term life?

Of course, once the general principles that govern such transformations have been characterized, there is still no guarantee that the historical question can then be resolved. After all, we are talking about particular events that took place on the earth some 4 billion years ago, so our ability to uncover the nature of those historical events is limited in the extreme. However, if and when that ahistorical question *is* resolved, the problem of how life on earth emerged on the prebiotic earth would take on a totally different aspect. Being a historical question the answer might remain unknown, but the issue would no longer be a mystery in the same way that it is now. Importantly, based on the above discussion, I am of the view that attempting to seek out life's molecular beginnings *before* we have adequately clarified the physicochemical principles that underlie biological complexification is tantamount to attempting to assemble a watch from its component parts—springs, cogs, wheels, etc.—without understanding the principles that govern watch function. Richard Feynman, the iconic Nobel physicist, once said: 'What I cannot create, I do not understand.' This truism might be usefully turned around: What I do not understand, I cannot create.

I have described in some detail the limitations in tackling the origin of life problem through its historical aspect, so let us now consider how the problem may be tackled through its ahistorical aspect. And it is here that we'll find room for greater optimism. Ahistorical principles are as relevant today as they were 4 billion years ago—the rules of physics and chemistry do not change over time. So rather than speculate as to what *might* have transpired on the prebiotic earth, let us investigate what *does* take place on today's earth. Let us study and experiment with chemical systems of the right kind, in order to glean information and obtain insight into this key question.

As we mentioned in chapter 4, systems chemistry deals with the class of simple replicating molecules and the networks that they create. That area of study, still in its infancy, has already revealed that reactivity patterns observed in such systems are quite different from those we find in 'regular' chemistry, and may provide insight into the kind of chemical processes that led to the emergence of life. In fact the switch in emphasis from historical to ahistorical leads us directly to an issue that has been central to the origin of life debate for several decades. Since all living systems are characterized by possessing a metabolism and the ability to reproduce themselves, which of these two capabilities came first—replication or metabolism? At first the question might sound historical in its approach—which came first? But the nature of these two capabilities may be such that chemical logic could dictate the natural order to be expected, and, as a consequence, could provide insight into the process of emergence. As we will see, the implications of the 'metabolism first—replication first' dichotomy are significant because they directly impact on all three questions that make up the triangle of holistic understanding, namely, what is life, how did it emerge, and how would one make it.

Before beginning the discussion let's make sure that the terms 'metabolism' and 'replication' are adequately defined. Broadly speaking the term 'metabolism' refers to the complex set of mutually regulated and coordinated reactions that take place within every living cell and which enable it to carry out life's processes. In the context of the origin of life question, 'metabolism first' mechanisms presume that some relatively simple autocatalytic chemical cycle, a forerunner of the complex metabolic cycles found in extant life, emerged prior to the appearance of an oligomer-based genomic system. As Stuart Kauffman, the influential theoretical biologist pointed out already in the 1980s, if within a set of molecules or molecular aggregates, say, A, B, C, D, and E, if A catalyses the formation of B, B catalyses that of C, C that of D, D that of E, and finally E that of A, then the closure of that cycle results in the entire cycle become autocatalytic, meaning that the system as a whole is self-replicating.[42] The 'replication first' school also views life as having been initiated by the emergence of an autocatalytic system, but in this case one based on a template-like oligomeric replicator, such as RNA (or RNA-like). Once such a replicator emerged it is then presumed to have evolved and complexified, eventually leading to the establishment of some simple life form. So the 'metabolism first—replication first' debate may also be expressed as which came first, the spontaneous formation of a holistically autocatalytic chemical cycle, or the emergence of some template molecular replicator?

Freeman Dyson, an American physicist, was the first to ask this question directly, and assumed that metabolic complexification and template replication are not logically connected. Dyson proposed that the origin of life involved the independent formation of two *separate* entities, one genomic, the other metabolic, which then combined to form a system that could be classified as alive, a system both genomic *and* metabolic.[43] That suggestion is actually quite arbitrary and, given the considerable scepticism with which the spontaneous emergence of *either* of those characteristics has subsequently been viewed, the likelihood of *both* characteristics emerging spontaneously and independently now seems highly unlikely. Consequently, the debate over the past decades has focused on the question (reductionist in its approach) as to which of these two special characteristics emerged first, molecular replication by a template mechanism, or holistic autocatalysis associated with a chemical cycle already exhibiting some level of complexity? *Does the essence of life derive from the sequential nature of certain oligomeric molecules, or from the complexity associated with holistic autocatalysis?* The fact that two schools of thought have emerged testifies most eloquently to the fact that neither school is compelling, each having its inherent weaknesses. The fact that the question is asked at all demonstrates too well how rudimentary our understanding of life continues to be. Let us begin by assessing the 'replication first' scenario in some detail and see why, despite its status as the basis for the widely-held RNA-world viewpoint, some fundamental difficulties remain unresolved.

'Replication first' scenario

As noted above, the 'replication first' scenario for the origin of life rests on the idea that life originated with the emergence of some oligomeric self-replicating entity and that replicating entity then proceeded to mutate and complexify until it became transformed into some minimal life form. Historically that idea can be traced back as far as 1914, to an American physicist, Leonard Troland, but that scenario was given a major boost through the contributions of Sol Spiegelman in the late 1960s that we described earlier. Within a short period of time those ideas were given further support through the pioneering works of Manfred Eigen and Peter Schuster in the 1970s.[44] Central to 'replication first' thinking was the proposal of an RNA-world that preceded the interdependent world of nucleic acids and proteins which forms the basis of all modern life.[45] A key attraction of the RNA-world scenario was that it appeared to solve the long-standing 'chicken and egg' dilemma with respect to the dual world of nucleic acid and protein. All modern life forms depend critically on this interdependence. DNA, the nucleic acid in which all heritable information is coded, cannot replicate without the elaborate involvement of protein enzymes, and those protein enzymes cannot be generated without the prior existence of the DNA molecule, which codes for those enzymes. So how could this dual world have come about? The RNA-world hypothesis appears to resolve this dilemma through its proposal that RNA originally functioned as both the carrier of genetic information *and* the provider of enzymatic activity. The fact that RNA can carry genetic information is not surprising. It is, after all, a nucleic acid closely related to DNA. But the discovery by two American researchers, Thomas Cech at the University of Colorado and Sidney Altman of Yale University, that RNA can also act as an enzyme and catalyse key biochemical reactions, gave the RNA-world viewpoint a major boost (as well as a Nobel prize to Cech and Altman). But the RNA-world view critically depends on the idea that a self-replicating molecule could have emerged spontaneously on the prebiotic earth, and that idea has continued to meet with opposition.

A central criticism of the 'replication first' scenario is based on the view that conditions on the prebiotic earth were not consistent with the spontaneous emergence of a molecule possessing a self-replicating capability. However, as discussed earlier, this view has no sound basis. The term 'prebiotic conditions', so frequently quoted in the origin of life literature, may convey some general information, but is totally devoid of specific information and so cannot be used to rule out any process, if that process is consistent with the basic rules of chemistry. Replicating molecules *can* be synthesized in the lab, so their spontaneous appearance on the prebiotic earth cannot just be dismissed *ad hoc*. Our ignorance regarding the prebiotic earth means that we cannot rule out the possibility that such an entity did in fact emerge on the prebiotic earth.

A more fundamental problem with the 'replication first' scenario is its apparent incompatibility with the Second Law of Thermodynamics. Let us recall what the 'replication first' scenario actually proposes. It rests on the idea that once some self-replicating entity happened to emerge, it then proceeded to complexify until it became transformed into some minimal life form. The difficulty with that proposal is that the simplest living system is a highly organized far-from-equilibrium system, which needs to constantly consume energy in order to maintain that far-from-equilibrium state. In other words for a replicating molecule to have complexified into a simple living system would have meant that instead of reacting to yield thermodynamically *more* stable products, it ended up becoming a highly complex thermodynamically *unstable* system. But that's not how chemical processes proceed. It's almost as if in a thermodynamic sense the reaction proceeded *uphill*, whereas, as we have seen, chemical reactions only proceed *downhill*.

So even if a replicating molecule *were* to emerge spontaneously, and even if it were to find itself in conditions that enabled the replication reaction to proceed, that reaction would only proceed until it reached the lowest free energy state, the equilibrium state. Once the system reached that low-energy state the process of evolution toward some minimal life form would cease. Indeed four decades of experimentation with replicating molecules has provided no indication of an inclination for such molecules to complexify toward far-from-equilibrium metabolic systems. For the 'replication first' scenario to be viable an explanation needs to be offered as to how a simple replicating system would be induced to complexify and 'climb uphill'. I will say more on this point subsequently. Let us now see how the alternative 'metabolism first' school of thought holds up to inspection.

'Metabolism first' scenario

A number of distinctly different mechanistic scenarios for the origin of life can be categorized as 'metabolism first' and we will not go into their details. The key point is that despite major differences in the essence of their chemistry, all contend that holistic autocatalysis (a catalytic cycle that achieves closure)—in what might be thought of as a primitive metabolism—preceded the subsequent incorporation of a genetic capability. Second, all presume that the organization required to generate metabolic function came about spontaneously, or through random drift. In other words the 'metabolism first' scenarios presume that the functional coherence inherent in metabolic processes can come about of its own accord, that *disorganized* systems underwent spontaneous *organization*. But, as has been pointed out by several leading origin of life researchers, in particular Shneior Lifson[46] and Leslie Orgel,[47] that idea is highly problematic. It's the Second Law problem again. How would metabolic cycles form spontaneously from simple molecular entities, and, more importantly, how would they maintain themselves over time? We run yet again into that thermodynamic brick wall. The same problem that puzzled physicists with respect to the emergence of *cellular* complexity is applicable to the emergence of *metabolic* complexity. Highly organized far-from-equilibrium chemical systems are not expected to be generated by spontaneous 'downhill' processes. And for those who say such transformations can take place, despite the Second Law, some experimental demonstration of such an occurrence is necessary. Harry Truman famously said: 'I'm from Missouri—show me.' So far no one has.

So both the 'metabolism first' and 'replication first' scenarios for the origin of life are problematic, not due to some minor issue, but because both have fundamental difficulties with the Second Law. We need to come up with a mechanism for the process of complexification toward a far-from-equilibrium system that does not contravene the Second Law. If, and when, that issue is resolved, the question of 'metabolism first' or 'replication first' may actually take on a different perspective. The answer to the question as to which came first may then become apparent, or, at the very least, may become less relevant. We will consider a possible resolution of this sticky problem in chapter 7.

Chance or necessity?

The prevailing view that life emerged from non-life leads to an immediate and highly problematic dilemma: was life's emergence on earth deterministic or was it contingent? In other words, was it a fantastically improbable accident—a freak occurrence that would almost certainly never be repeated—or was life's emergence inevitable given the existing laws of physics and chemistry. Two Nobel prize-winning biologists have famously faced off on this question. Jacques Monod viewed it as a bizarre accident unlikely to be repeated.[10] In his words: 'That would mean that its *a priori* probability was virtually zero... The universe was not pregnant with life nor the biosphere with man.' Christian de Duve, however, takes the opposite view and considers the emergence of life on earth-like planets a 'cosmic imperative' governed by the laws of chemistry and physics.[48] De Duve goes as far as to contradict Monod with the statement: 'It is self-evident that the universe was pregnant with life and the biosphere with man. Otherwise, we would not be here.' So who is right? Did life on earth emerge by chance or necessity (to paraphrase the title of Jacques Monod's classic text)?

The first point to note is that the Monod and de Duve positions are actually extreme ends of a continuous spectrum of possibilities. To illustrate this point consider the probability of snowfall during winter. Is snowfall in winter deterministic or contingent? In the Swiss Alps snowfall during winter would be considered deterministic. Due to the low temperatures that prevail in the Alps in winter the probability of snowfall is extremely high. Pretty well guaranteed. But on a Queensland beach the probability of snow falling, even in winter, is very close to zero. Queensland temperatures don't get low enough. What about snowfall in Rome? Here the probability is intermediate—it does snow in Rome on occasion. In the last thirty years it snowed in 2012, 2005, and in 1986. Snowfall in Rome is a contingent event. The conclusion? A particular event could in principle be highly contingent or effectively deterministic or anywhere in between.

Of course one doesn't have to understand the physics of snowing to be able to state whether snowfall at a particular location is deterministic or contingent. Simply by checking the historical record regarding snowfall at that location, you will have the answer. That's why we can be supremely confident it will snow in the Alps this winter and that it won't snow on the beach in Queensland. Regarding Rome, we must remain uncertain. All one can say definitively is that it *may* snow next winter in Rome, the probability being something like 10 per cent.

So, what can we conclude regarding the emergence of life on our planet? The short answer: almost nothing, and there are several reasons for that frustrating state of ignorance. In contrast to the meteorological phenomenon of snowfall which is well understood, we don't understand the process by which life emerged, and we are relatively ignorant regarding the prevailing conditions at the time. How can one expect to be able to judge the likelihood of a process we don't understand and which took place under unknown conditions? Alternatively, as in the case of snowfall, one might be able to make a prediction without understanding the process, simply by carrying out a historical survey of the phenomenon in question. But here we run into a different problem. Our survey is restricted to a sample of one. Even though we are aware that the number of earth-like planets in the universe is likely to be spectacularly large, we only know the life situation on one of these—our own. With a sample of just one to guide us, our ability to reach a reasoned assessment of its likelihood elsewhere in the universe is obviously limited.

6

Biology's Crisis of Identity

The difficulties in relating living and non-living entities, first with respect to the very strange characteristics of living things (chapter 1) and then with regard to the seemingly intractable origin of life problem (chapter 5) have exposed the scientific quandary that modern biology has been contending with in recent years. In fact three core questions at the heart of the subject—what is life, how did it emerge, and how would one make it—remain troublingly unresolved. And though these questions may initially seem independent and quite unrelated, they are in fact intimately interconnected, as schematically illustrated in Fig. 5. If you think about it, being able to answer any one of the questions depends on knowing the answers to the other two. We don't know how to go about making life because we don't really know what life is, and we don't know what life is, because we don't understand the principles that led to its emergence. So, despite those spectacular advances in molecular biology over the past sixty years, the very essence of what biology claims to study remains troublingly obscure. That

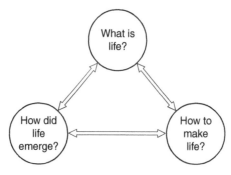

Fig. 5. Three key questions governing holistic understanding in biology

gloomy view is not just the frivolous opinion of an over-zealous chemist on a subject that is not his own, but one that is beginning to be expressed more generally. Carl Woese, in an almost messianic article that we have already referred to, recently wrote:[1]

> Biology today is no more fully understood in principle than physics was a century or so ago. In both cases the guiding vision has (or had) reached its end, and in both, a new, deeper, more invigorating representation of reality is (or was) called for... Look back a hundred years. Didn't a similar sense of a science coming to completion pervade physics at the 19th century's end—the big problems were all solved; from here on out it was just a matter of working out the details? Déjà vu!

Woese, a leading contributor to the molecular approach to biology whose fruits have been so rewarding, seems to have lost all faith in the methodology that served him and molecular biology so well. Paradoxically it is the dramatic increase in knowledge brought about by molecular biology that has actually revealed how ignorant we are. So what went wrong?

The road from Darwin to modern biology was a convoluted one. Darwin's monumental achievement was, of course, in providing

BIOLOGY'S CRISIS OF IDENTITY

biology with a physical foundation, thereby successfully transplanting biology from the supernatural world into the natural world. In doing so, Darwin irrevocably changed our perception of ourselves and the world in which we live. But it was far from smooth sailing. First, natural selection, the very heart of Darwinism, was not fully accepted by biologists till well into the twentieth century. It was almost eighty years after the publication of *Origin of Species*, in the 1930s, that Darwinian theory was finally embraced, as part of what is termed the modern evolutionary synthesis. It was the winning integration of Darwinian evolutionary theory with Mendelian and population genetics that finally eliminated academic doubts as to the significance of the Darwinian legacy. That integration provided the mechanism by which natural selection could perform its magic, thereby eliminating the main sources of prevailing criticism.

But another revolution was beginning to build up momentum—the revolution in molecular biology. Indeed as already noted, a half-century of dramatic discoveries beginning with the structural elucidation of DNA in 1953 were revealed in quick succession—DNA replication, RNA transcription, protein translation, the ribosomal machine, with a long string of Nobel prizes illuminating the path to what Walter Gilbert termed the Holy Grail—elucidating the entire base sequence of the 3 billion bases in human DNA, the human genome project in which the entire human genome was sequenced. The reductionist dream appeared to have been realized, the essence of humankind had been reduced to a string of 3 billion letters. On its completion in 2000, Bill Clinton in a White House ceremony dramatically claimed 'today we are learning the language in which God created life' and added that the achievement would 'revolutionize the diagnosis, prevention and treatment of most, if not all human diseases'. Personalized medicine was promised by 2010. A brave new world was with us, the mysteries of biology were finally solved. Any lingering details still to be resolved were just that—details, hardly worth mentioning in the big scheme of things. Just the way physics felt at the end of the nineteenth century...

Well, at the time of writing, the so-called Holy Grail and the language of life that it was supposed to have taught us have not delivered the promised rewards. Not only hasn't early twenty-first-century biology reached its goal of solving the major biological problems, but there is a growing awareness that there is a largish elephant in the room. Life is more complicated than a representation provided by a string of 3 billion letters. The gap between the elucidation of the human genome sequence and understanding the significance of that sequence is cavernous. The uncovering of more and more structural and mechanistic information within the living cell hasn't clarified what life actually is. Stuart Kauffman[42] put it in succinctly in his thought-provoking text *Investigations*:

> despite the fine work...in the past three decades of molecular biology, the core of life itself remains shrouded from view. We know chunks of molecular machinery, metabolic pathways, means of membrane biosynthesis—we know many of the parts and many of the processes. But what makes a cell alive is still not clear to us. The center is still mysterious.

What both Kauffman and Woese were effectively saying, each in their own words, was: we see so many trees, yet we have no real view of the forest.

So where's the problem? The answer in a nutshell is complexity, the organizational complexity that is life. The reductionist strategy for dealing with complexity seems to have floundered. It works great for clocks, it has been a boon for our understanding of the natural world, but its performance in the life arena has been mixed. The spectacular advances in molecular biology, reductionist in its approach, have not opened the gates to the Promised Land. Our attempts to view biological systems as mechanical-materialistic machines have failed dismally. The reductionist methodology has not as yet brought us any closer to answering the basic life questions depicted in Fig. 5, nor the global ones that we discussed in detail in chapter 1. Tibor Ganti, a Hungarian chemical engineer, recognized the problem over thirty-five years ago when he stated that 'living systems have special properties which arise primarily not from the substances of the system, but from their special organizational manner.'[49] It is the *organization* of life rather than the *stuff* of life that makes life the unique phenomenon that it is.

So how do we deal with that troublesome issue of organization—the very special kind of organization associated with all living things. Over recent decades, scientists of various kinds—physicists, chemists, mathematicians, and not just biologists, have been exploring alternative approaches to the problem. One direction taken was to argue that physics and biology were necessarily based on different philosophies of science and, therefore, that biology should be recognized and treated as such. For fundamental but unspecified reasons, biology was viewed as not being reducible to physics, even though reduction had proven so successful in bridging between physics and chemistry and making sense of the inanimate world. Divide and conquer! Indeed, that way of thinking seems to have bolstered the holistic approach to biology, as expressed by the burgeoning area of systems biology. Let me briefly describe that approach, both its benefits and its apparent shortcomings.

In contrast to molecular biology, in which the focus is on the structure and reactivity of individual molecules and molecular aggregates within the cell, systems biology attempts to address the manner in which these cellular components interact as a system. After all it is the system as a whole, not just individual components, that is responsible for biological function. Inherent in the systems approach is the belief that there are general features of systems, in particular their network topology, which characterize the system's behaviour, and that such understanding may provide biological insights.

But the systems biology approach has not proved a nirvana. General rules governing the behaviour of complex systems have not as yet been delineated and in any case it is clear that without sufficient attention to component functionality, the systems insight will on its own be necessarily limited. As we discussed in chapter 3, the terms reduction and holism do not have to be mutually exclusive, so that in many respects the holistic approach can be thought of as reductionism dressed up. Given this inconclusive and rather unsatisfying situation, systems biology has routinely resorted to falling back on the concept of 'emergent properties' whenever the system's properties are not readily explained through a reductionist approach. But the use of that catchphrase is a double-edged sword. Sweeping unexplained phenomena under the expansive complexity carpet creates the illusion of explanation, which, in itself, can be problematic. A phenomenon that is unexplained will continue to attract attention until some convincing explanation is offered. But once some unexplained phenomenon is classified as an 'emergent property', it could be thought of as explained, that no further consideration of the phenomenon is required. How else to understand the almost total lack of interest that the scientific community has shown in the physicochemical basis for teleonomy, that most remarkable of emergent properties? Jacques Monod in his classic *Chance and Necessity* considered the problem of teleonomy as the 'central problem of biology'.[10] As Monod put it: how could purposeful systems have emerged from a universe with no purpose? But the minimal attention that has been directed toward this 'central problem' suggests that the scientific community considers the problem solved (or uninteresting) and has accepted the 'emergent property' explanation.

Another reason that the biological community might have ignored Monod's challenge is that his question might sound more philosophic than scientific. But don't be fooled. The question of how purpose and function can manifest themselves spontaneously is a profoundly important scientific question and its resolution would help connect chemistry, representing the objective material world, with biology, representing the teleonomic world. Bottom line: Darwinism did bring about a sense of unity *within* biology, but the troubling consequence of that unification, of enormous value in itself, has been a growing isolation of the subject from the physical sciences to which it must necessarily connect.

Let us now briefly consider two other approaches that have been taken in the past several decades to try to crack the complexity nut—one physical, the other mathematical, and review their current status. The physical approach to the problem came about through the observation of physical systems, such as hurricanes, whirlpools, vortices, and the like. The theory of such systems, attributed primarily to work of the Belgian physical chemist Ilya Prigogine in the

1950s and 1960s is covered by what is termed non-equilibrium thermodynamics[50]—a mouthful for those who do not work in the area, and I will spare the reader a detailed discussion. The main point worthy of mention is the fact that some connection between certain non-equilibrium physical systems and biological systems is claimed to have been found. Recall that one of the mysteries of biological systems is how their non-equilibrium complexity came about naturally. But fill a bath tub with water, pull out the plug, and from a purely physical point of view something intriguing occurs. Whereas the body of water in the bath is in a stable state when the plug is in place, removing that plug creates an *unstable* situation as the body of water can reduce its potential energy by flowing down the drain. Of course the body of water immediately responds to this unstable situation—it begins to flow down the drain in order to lower its potential energy and reach a new equilibrium state. But in doing so something special takes place—the body of water generates a structure, a vortex. The *non-equilibrium state* has spontaneously generated a *non-equilibrium structure*. The body of water that initially lacked any structure has in some sense acquired order. In the language of non-equilibrium thermodynamics that structural pattern, which is evidenced in other physical systems as well, is termed a *dissipative structure*.

This physical pattern has led to the idea that in purely energetic terms some similarity exists between dissipative structures and living cells. Both are non-equilibrium, meaning they are unstable, and both have generated a non-equilibrium structure which must continually consume energy to maintain itself (in the case of the bath the source of energy is the lowering of the water's potential energy as it flows out of the bath). In other words the claim is made that one of the mysteries of life may have a simple physical resolution. Organization *can* be induced in an open system that consumes energy. The far-from-equilibrium organization of a living cell may in some sense be thought of as mimicking the non-equilibrium order induced in a tub of water or a heated column of liquid. The mystery of biological organization may have been at least partly resolved. These ideas were discussed with some enthusiasm some 20–30 years ago and without going into further detail, the approach seems to have lost much of its earlier appeal. The main difficulty is that the disarmingly simplistic connection between the physical and biological systems mentioned above did not lead to any useful biological insights. A model is only useful if it provides new insights and makes novel predictions. However, as was noted some years ago by John Collier, a philosopher at the University of Calgary, there is no evidence that the laws of non-equilibrium thermodynamics apply to biological systems in a *non-trivial* fashion.[51] Non-equilibrium thermodynamics has not proved to be the hoped-for breakthrough in seeking greater understanding of biological complexity. A physically based theory of life continues to elude us.

Enter the mathematical approach to complexity. In 1970, the Princeton mathematician John Conway invented a game which he called Life, which leads to interesting insights.[52] The game is based on a two-dimensional square grid where each square can exist in one of two states, dead or alive, most simply represented by the squares being black (alive) or white (dead). One starts the game with some particular limited pattern of live squares and then based on a rule that is specified, all eight squares surrounding each square (neighbours) are then made black or white, depending on the particular rule chosen. For example, the rule may be that any live cell with fewer than two live neighbours dies (it becomes white), any live cell with two or three live neighbours stays alive (stays black), any live cell with more than three live neighbours dies (becomes white), and any dead cell with exactly three live neighbours becomes a live cell. The process then is repeated many times to see how the initial pattern evolves over time. Depending on the starting pattern and the rule of the game, very different patterns can emerge. Sometimes the pattern remains unchanged (for example if the above rules are applied to a starting pattern of a 2×2 square block of live squares), sometimes it simply disappears after a few runs, but sometimes, quite extraordinarily complex patterns result. The game of Life teaches us that simple rules can lead to quite complex patterns and while the rules specified in the game of Life have no relevance to real life, the fact that complex systems may result from the operation of relatively simple rules is informative in itself. In fact we will demonstrate in the next chapter that real life does indeed appear to be governed by a simple rule, though we will need to discuss the nature of simple replicating systems before that rule can be appreciated. While Conway's Life game has opened up interesting insights into complex systems in general, direct insights into the nature of living systems do not appear to have been forthcoming.

Let us sum up the key conclusions from the above discussion. We have already noted that the problem of 'understanding life' involves more than merely accumulating further molecular insights into life's mechanisms. As the younger generation might say: been there, done that. We need to be able to explain life's complexity and the *global* characteristics associated with that complexity, and we are far from being able to do that. The non-equilibrium thermodynamic approach discussed above, though interesting in its own right, appears to have led to a dead-end. With regard to the attempts of biologists to better understand life's complexity through the newly emergent area of systems biology, the jury is still out. But current indications are that no major breakthroughs are imminent. While a systems biology approach may provide insights into specific biological problems, there is no indication that the approach is able to resolve the larger questions that have been raised. And though the mathematical approach to complexity has been instructive in offering new insights into complexity in general, it does not appear to have contributed in any significant way to untangling the tangled web that is particular to biological complexity. So how to proceed? In the final two chapters we will attempt to show how recent fascinating results within a newly founded and burgeoning area of chemistry can finally provide some concrete answers.

7

Biology is Chemistry

Systems chemistry to the rescue

Our earlier discussion has identified the nature of biological complexity as *the* nut that needs to be cracked. So, in addressing this problem, has reduction, that tested scientific methodology that stood us in such good stead these past several hundred years, reached its effective limits? Is a new methodological approach needed? A range of prominent biologists have been arguing in the affirmative. However, my answer remains *no*. In this chapter, I will describe the basis of that view, and attempt to demonstrate that there *is* a way forward, that the reductionist approach *can* be effectively applied to biology at the global holistic level. I will attempt to show that the chasm separating biology and chemistry is bridgeable, that Darwinian theory can be integrated into a more general chemical theory of matter, and that biology *is* just chemistry, or to be more precise, a sub-branch of chemistry—replicative chemistry. Despite the widespread concerns that have been raised with regard to the reductionist methodology in biology, the organizational issue *can* be resolved through a reductionist analysis.

In chapter 4, I mentioned that a relatively new area of chemistry, systems chemistry, has taken shape in recent years. This new field came about in trying to seek out the chemical origins of biological organization, and that explains its name, a play on words with its better-known cousin, systems biology. If we think of biology as the field of endeavour that studies those highly complex chemical systems capable of replication or reproduction, then systems chemistry (or at least central aspects of it) deals with relatively *simple* chemical systems that also possess that special characteristic of self-replication, and in doing so attempts to fill the chasm-like void that continues to separate chemistry and biology. In contrast to systems biology, which takes a 'top-down' approach in its attempt to contend with life's complexity, systems chemistry takes a 'bottom-up' approach. A top-down approach starts with what we have and works down from there seeking to understand the manner in which the components contribute to the whole. A bottom-up approach, needless to say, goes the other way—it starts from some presumed beginning and works its way up. In the life context that means that life's complexity is addressed by investigating the manner in which complexity was built up, step by step, from some initial simple entity, from the bottom up. A key challenge of systems chemistry then becomes to ascertain the rules, if such rules exist, which govern that process of complexification from a relatively simple chemical system to the highly complex systems that define present-day biology.

There are a number of factors that argue favourably for the bottom-up approach. First, we have already noted that life is presumed to have had its beginnings in inanimate matter, i.e., life emerged from non-life. That being the case it necessarily follows that life's beginnings *were* simple and that its complexity was built up over an extended time period, step by step. That, in itself, confers on the bottom-up approach a crucial advantage. The path leading from bottom to top is not merely conceptual—a *gedanken* experiment—but an *actual* pathway that was followed by a real chemical system. It now seems increasingly likely that several billion years ago some replicating system of unknown identity, but of low complexity, set off along the long and winding road toward high complexity, and that historic path of ever-increasing complexity eventually led from the world of chemistry to the world of biology. The fact that a reasonably well-defined process of complexification can be identified suggests that there may well be a *driving force* for that process, and one of our goals will be to seek its identity and explore its nature. Can that process of complexification be understood in physical terms?

Second, it seems logical to suggest that if life did start off simple, then life's fundamental nature would become more understandable by examining earlier, and therefore *simpler* prototypes. An analogy may make this clear. If we want to understand what an airplane is, as well as the underlying principles that enable these modern behemoths to take to the air, then examining a fully equipped Boeing 747 will not be the most productive way to proceed. A Boeing 747 is an immensely complicated entity composed of some 6 million individual parts and over 200 kilometres of wiring, so figuring out the relevance of each and every part to the whole, and uncovering the basis for its flying capability, would be overwhelmingly difficult. Some of those parts, for example, passenger TV screens, steward call buttons, ovens for heating food, etc., wouldn't be particularly relevant to its flying capability. So where is one to begin? If you want to figure out what an airplane is, and the principles governing its flight, you'd be much better off examining an earlier simpler airplane, say the Wright brothers' 1903 prototype or some other simple equivalent, where the number of components is a tiny fraction of that in the Boeing, and one in which every component plays an important, if not critical role in enabling that entity to become airborne. And that's where systems chemistry comes in—by examining the workings of simple replicating systems and the networks they generate, we are attempting to do the equivalent of examining the Wright brothers' airplane rather than a Boeing 747.

Of course the bottom-up approach toward resolving the life issue assumes that life *did* start off from simple beginnings and that a process of complexification from that simple beginning did take place. As discussed in chapter 5, that is the generally accepted view. It is the *nature* of that process that continues to be a source of intense debate, rather than whether the process took place. But, as we will shortly see, the emergent area of systems chemistry will also provide additional empirical support for that assumption. The goal of this chapter is therefore ambitious: to demonstrate that the study of systems chemistry can lead to the smooth merging of living and non-living systems, thereby offering a unifying framework for chemistry and biology. Such unification would be of considerable value as it would place biology within a broader chemical context. Indeed, if successful, that endeavour could provide fundamental insight into the 'what is life' question as it could offer a description of living systems in *chemical* rather than *biological* terms. So despite recent misgivings regarding the reductionist methodology as applied to biological systems, we will attempt to show that reduction in biology

Fig. 6. Two-phase (chemical and biological) transformation of non-life into complex life.

is alive and kicking (no pun intended!). In addition, a not insignificant side benefit would be to demonstrate that systems chemistry can throw light on the origin of life problem, at least in an *ahistorical* sense, by uncovering the *principles* that would have enabled inanimate matter to complexify in the biological direction toward life.

Let us then begin our discussion with the traditional view for the transformation of non-life into complex life. This can be represented as a two-stage process as illustrated in Fig. 6.

The first stage, the so-called chemical phase (termed abiogenesis, meaning the process by which life emerged from non-life) is where the never-ending debate and controversy lie. In the context of Fig. 6, a simple life form would mean that the system would possess what many would argue would be the most significant characteristic of living things—the ability to replicate and evolve in a self-sustained way. Indeed, having reached that critical point, the system would be considered biological in nature and its subsequent transformation into more complex life—single-celled eukaryotes and multicellular organisms—would have been governed by that momentous and earth-shattering theory that was proposed just 150 years ago, Darwinian evolution. So the conventional wisdom is that we are facing a two-stage process whose first stage is highly contentious and uncertain, while the second stage, in scientific circles at least, is in broad terms now unshakeable.

Let me now drop the bombshell, at least for many in the field. The so-called two-stage process is not two-stage at all. It is really just *one single continuous process*. If true that statement has quite profound consequences. First it must mean that hidden within Darwin's theory of evolution—biological in formulation and application—a more fundamental, broader principle is at work, which must necessarily incorporate prebiotic systems, which by definition would be classified as non-living. In this chapter, I will attempt to justify the one-process assumption and explore some of its implications.

Why has the process indicated in Fig. 6 been considered a two-phase process until now? To put it bluntly—because of our ignorance. Knowing the mechanism of one phase and not knowing the mechanism of the other is a clear point of division and leads quite naturally to the separate classification. However, ignorance is not a useful basis for classification, so let me now try to justify the assertion that abiogenesis and biological evolution are in fact one single continuous process. And I don't mean that in a trivial sense. It is obvious that if some prebiotic entity complexified into a simple living thing by some unknown mechanism, and then proceeded to evolve and diversify into the extraordinary range of living species, then whatever that early prebiotic process was, it could be thought of as continuous with the biological phase, at least in a temporal sense. But I intend the statement in a *non-trivial* sense—that the chemical process that led to the simple living creature and the biological process that subsequently carried on from there are one single process in a *chemical* sense. That is in fact exactly what recent studies in systems chemistry have been telling us. Let us review the empirical evidence.

In chapter 4, I described the molecular replication reaction of an RNA molecule carried out by Sol Spiegelman in the 1960s. We saw that molecular replication is a chemical reality and can take place in a test tube, and not just in the highly regulated and specific environment of a cell. Recall, however, that Spiegelman also discovered that the population of replicating RNA molecules can evolve.[27] Over time the initially long chain RNA molecule evolved into shorter RNA chains. Shorter RNA molecules which replicated faster, out-replicated the longer ones, driving those longer ones to extinction. So what is termed natural selection within the biological world is also found to operate in the chemical world. The conclusion is highly significant. The causal sequence: replication—mutation—selection—evolution, normally associated with the biological world, in fact the *sine qua non* of biology, is also clearly evident at the chemical level. That landmark work was carried out over forty years ago and since then the phenomenon of molecular evolution—evolutionary-like behaviour at the molecular level—has been observed by a growing number of researchers. Accordingly, the generality of evolutionary processes within replicating entities at the molecular level is now well documented and experimentally uncontroversial.

But the chemistry-biology nexus runs much deeper. Ecology is an established branch of biology and would seem to be quite unrelated to chemistry. However, as Gerald Joyce, the remarkable Scripps biochemist, reported in 2009, there is an intimate connection between the two.[53] A key ecological principle, termed the *competitive exclusion principle*, states: 'Complete competitors cannot exist' or, expressed in its positive form: 'Ecological differentiation is the necessary condition for coexistence'.[54] What that principle teaches us is that two non-interbreeding species that occupy the same ecological niche (which just means that the two species compete for the same resources) cannot coexist—the one that is better adapted to that niche (i.e., is fitter), will drive the other to extinction. Of course, if the two species feed off *different* resources then coexistence *is* possible. This basic ecological principle is classically illustrated by Darwin's finches—one of the best-known examples of evolutionary theory. On the Galapagos islands, where Darwin visited in 1835, one can find a variety of finches that differ in the size and shape of their beaks. These different finch varieties, all of which derive from a common ancestor, evolved over time so as to exploit available resources more effectively. In doing so one type of finch—ground finches—evolved strong beaks which are effective for cracking nuts and seeds, while another type—tree finches—evolved sharp pointed beaks which are adapted for eating insects. The point is that these distinct varieties of finch can coexist because each is adapted to feed off a different resource, and in that sense provide a good example of the competitive exclusion principle.

But here is where the chemistry-biology connection comes in. Gerald Joyce discovered that this quintessentially biological principle also operates in chemistry.[53] Joyce found that when two different RNA molecules, let's call them RNA-1 and RNA-2, were allowed to replicate and evolve in the presence of some essential substrate they were unable to coexist. RNA-1 turned out to be the more effective replicator with that substrate, and as a consequence it drove RNA-2 to extinction. If a different substrate was employed, one that RNA-2 was able to exploit more effectively, then the result was reversed. Now it was RNA-1 that was driven to extinction, as RNA-2 was the more effective replicator in the presence of that other substrate. Those chemical results are precisely in line with the predictions of the biological competitive exclusion principle. Since both replicators relied on the presence of a particular substrate in order to replicate, the two molecules were unable to coexist—the faster (fitter) replicator drove the slower one to extinction.

But a more interesting and quite remarkable result was to follow: when the two RNA molecules were allowed to replicate and evolve in the presence of not *one*, but *five* different substrates, the two RNAs *were* able to coexist, but in an unexpected way. Initially the two RNA molecules utilized all five substrates in varying degrees in order to replicate. After all, all five were present and therefore all five could be utilized to some extent. But here is the punch line: over time each RNA molecule evolved so as to optimize its replicative ability with respect to *different* substrates. RNA-1 evolved so as to optimize its replicative ability with just *one* of the five substrates, while RNA-2 evolved so as to optimize its replicative ability with another of those five substrates. As a result, the two RNAs *were* now able to coexist.

In this beautifully designed experiment, which explored the characteristics of competing molecular replicators, the two RNA molecules were found to mimic the behaviour of Darwin's finches precisely! Each molecule evolved to exploit a particular substrate efficiently, just as Darwin's finches had evolved beak size and shape to suit the nature of the resource. That spectacular result, in which molecular replicators mimic biological ones (actually vice versa—molecular replicators preceded the biological ones), speaks loud and clear for a strong chemical-biological connection. Darwin's finches are merely doing what certain molecules started doing billions of years ago.

Finally, let me show that complexification of the special kind normally found only in biological systems can also be discerned at the chemical level, and so provides yet another link between

BIOLOGY IS CHEMISTRY

chemical and biological replication processes. We have already discussed the fact that complexity is the very essence of biology. In fact, over an evolutionary time frame it is quite evident that complexity has continually increased from relatively simple systems to more complex ones. The earliest life forms that emerged, perhaps 4 billion years ago, were simple cells, prokaryotes (meaning that the cells lack a nucleus and other organelles). But after a further 2 billion years of evolution, eukaryotic cells emerged, in which membrane-bound organelles, including the cell nucleus, can be found. And some 600 million years ago another evolutionary transition involving further complexification took place, the one in which multicell organisms—plants and animals—appeared.[55] The evidence on this score is therefore unambiguous. Over the evolutionary time frame there has been a clear tendency for complexity to increase (though of course only among a small subsection of life, the multicellular eukaryotes; the vast majority of life, bacteria and archaeans, have remained happily simple). So within what we have labelled as the biological phase of Fig. 6, there is unambiguous evidence for a process of increasing complexity.

What can we say about the chemical phase of Fig. 6? In historical detail, almost nothing at all. But the essence of the transformation is quite clear. A molecular system, which we would characterize as non-living and relatively simple, somehow became transformed into a highly complex living cell, meaning the process involved was one of increasing complexity. As we have already pointed out, even the simplest living thing is highly complex. In other words, *both chemical and biological phases of Fig. 6 involved a process of continual complexification.* But how can this process of apparent complexification be understood at the chemical level?

As we have discussed in some detail in chapter 5, we are lacking any direct information regarding that early prebiotic period. However there is one thing we can state with high assurance regarding that early period. It is that the laws that govern chemical behaviour have not changed over the past several billion years, and that means that studying the right kind of chemistry today can inform us about what might have happened billions of years ago. And the right kind of chemistry is systems chemistry, the chemical reactions of replicating molecules and the networks they create.[29,56] Such study may provide us with insight into the *kinds* of reactions that prebiotic replicators might have undertaken, amongst them that early process of complexification.

What have we then learnt regarding simple chemical replicators? First, getting single molecules to self-replicate is inherently difficult. In fact the difficulty in getting so-called replicating molecules to replicate when no biological materials are added to 'help' things move along, has been viewed as one of the stronger arguments against a replication-first scenario for the emergence of life. But let us return to some recent results from the Joyce lab as they are illuminating. Despite the difficulties inherent in getting single molecules to replicate, Gerald Joyce was able to come up with an RNA molecule that was able to make copies of itself without enzymatic assistance. In that particular reaction, a replicating RNA molecule, let's call it T, itself composed of two RNA segments, A and B, underwent a replication reaction by the template mechanism (described in detail in chapter 4). The RNA molecule, T, acting as a template, induced fragment entities, A and B, which were floating about freely, to temporarily bind to itself and then link up, thereby creating a new molecule of T. The overall result, a single T molecule was able to make copies of itself by inducing its two component parts A and B, freely available in the solution, to connect up.[57]

Even though that replication reaction was possible, it was frustratingly inefficient. First, it was slow—it required seventeen hours for an initial sample of RNA to double in quantity. But slowness wasn't the only problem. After all what is seventeen hours when compared to a billion-year time frame? An additional problem was that the replication reaction only proceeded for two replication rounds before grinding to a halt (due to certain side reactions), so it was not possible to continue the reaction, even when feedstock for further replication reactions (i.e., more A and B) was provided. But now to the interesting finding. When Joyce switched from a single replicating RNA molecule to a two-molecule system composed of *two* discrete RNA molecules that had been obtained in a careful selection process, then replication proceeded rapidly—the initial sample doubled in quantity in just one hour—and replication could be sustained indefinitely, provided the building blocks were available. How come? Why the difference?

Let's start by stating what wasn't happening. In the two-molecule RNA system each molecule was *not* making copies of itself. Rather, one RNA molecule was inducing the formation of the other, while the other molecule was inducing the formation of the first. In chemistry we call that cross-catalysis—each RNA molecule was catalysing the formation of the other. So the more complex system *is* self-replicating, but in a more complex way—each component of the system isn't replicating individually, but the system as a whole *is* self-replicating. That distinction is important because holistic replication is the norm in biology; that's what cells do when they replicate—the system as a whole makes copies of itself, as opposed to each individual component within the cell copying itself. So what is the significance of this result? Simply this: what one simple replicating entity could only do *inefficiently*, a more complex one was able to do *more efficiently*.

This chemical equivalent of 'I'll scratch your back, if you'll scratch mine' goes beyond the tit-for-tat exchange of favours, which is useful in itself. The deeper meaning is that what I cannot do well on my own, I can do more effectively in a cooperative way. Cooperation is win-win. No wonder cooperation is endemic in the biological world—biologists call it symbiosis. You see it wherever you look. So what Gerald Joyce discovered in those two RNA molecules was profound. Yet another piece of evidence that demonstrates how chemistry and biology are intimately connected. A process of *molecular complexification* has led to an enhanced replicative capability.

Let us take another look at Fig. 6 because it now takes on a new significance. Our discussion above has indicated that complexification facilitates both the molecular replication phase *and* the biological replication phase. In fact, the entire process when viewed over an evolutionary time frame is seen to be one of complexification. The main difference between the two phases is that the first phase, the chemical phase, is the *low-complexity* phase, while the second phase, the so-called biological phase is the *high-complexity* phase, all taking place within the context of replicating entities. The conclusion seems clear: complexification, primarily through network establishment, appears to be the *mechanism* for the transformation of *simpler chemical* replicators into more *complex biological* ones. In fact the recognition that complexification is a key process in evolution leads us to a surprising conclusion, namely, that the

BIOLOGY IS CHEMISTRY

causal sequence that leads to evolution needs to be modified. Evolution in biology is normally associated with the causal sequence: *replication, mutation, selection, evolution*. But we now see that an important step in that sequence has been overlooked. The missing step is complexification. The sequence should read: *replication, mutation, complexification, selection, evolution* and this is true for both the chemical and biological phases.

Some words of clarification are now appropriate. The previous discussion might suggest that the evolutionary process is based solely on complexification and this is clearly not the case. It is well established that in particular instances evolution follows a process of *simplification*. Biology in particular is replete with such cases—for example, cave-dwelling animals such as crickets and cavefish that lose their eyesight as they adapt to life in the dark. But, remarkably, in chemical systems precisely the same phenomenon of simplification can also be observed. Recall Spiegelman's experiments on molecular evolution in which replicating RNA chains shortened because the shorter chains replicated faster.[27] That classic study provides a *chemical* example of simplification. Just as cavefish lose their ability to see in dark caves, RNA chains (extracted by Spiegelman from the Qβ virus) discard those parts of the viral genome that prove redundant in the artificial resource-rich test-tube environment. The very existence of a process of simplification in both biological *and* chemical evolution serves to *further* strengthen the chemistry-biology connection and provides yet an added piece of evidence supporting the unity of the evolutionary process of Fig. 6. Returning however to the present theme, regardless of those well-documented instances of simplification, it is clear that complexification is the *underlying* tendency in evolution, in both the chemical and biological worlds.

In the light of the above experiments and arguments, the reader has hopefully been convinced that the processes of abiogenesis and evolution are actually one single physicochemical process governed by one single mechanism, rather than two discrete processes governed by two different mechanisms. That insight will turn out to be of utmost value as it leads to a whole range of both chemical and biological insights. If our conclusion is correct it means we can apply *chemical* insights from the chemical phase to better understand the *biological* phase, and we can also apply *biological* insights derived from 150 years of studying Darwinian evolution to provide greater insights into the *chemical* phase. Win-win for sure! But beyond that, the unification tells us that chemistry and biology are one, that there is a *complexity continuum* that connects them, that biology is just an elaborate extension of replicative chemistry. Interestingly, as noted in the prologue, Darwin, in his genius, foresaw the existence of some underlying principle governing abiogenesis and biological evolution. However, thanks to the inspiring work of gifted systems chemists these past decades, we don't have to speculate about the nature of a general life principle—the life principle can now be formulated based on hard facts.

So what new insights does this merging of chemistry and biology provide us with? Before answering this question and in order to fully benefit from this conceptual merging, we now need to rephrase Fig. 6. Traditionally one would describe the first phase, the chemical one, in chemical terms, and the second biological phase in biological terms—each process in its own language. But, as we all know from foreign travel, a dialogue in two languages, when the two parties do not speak the other's tongue, may be frustratingly less than useful. Misunderstandings galore can arise. In order to avail ourselves of the deeper insight of one continuous process, the two phases need to be described in one language. So which is it to be—the language of chemistry, or that of biology? The answer is clear-cut: the entire process—chemical and biological—needs to be described in chemical terms. Let me explain why.

In an earlier chapter (chapter 3), I described how understanding in science is achieved at different hierarchical levels. Phenomena at a higher hierarchical level of complexity are normally explained in terms of scientific principles associated with a lower hierarchical level. Thus we conventionally explain biological phenomena in chemical terms and chemical phenomena in physical terms, not the other way around. Recall, Steven Weinberg's comment: 'Explanatory arrows always point downward.'[24] To bring this point home and to illustrate how fundamental this hierarchical aspect of explanation is, consider the two sciences, chemistry and psychology, and how they might interrelate. Let us say you find some psychological phenomenon of interest and you tried to explain it in molecular terms. Scientifically speaking that is quite acceptable. For example, if you came up with a molecular explanation for schizophrenia that would certainly be of interest—drug companies would likely be knocking on your door! However, if one went the other way and attempted to explain some *molecular* phenomenon in *psychological* terms, that would only attract derision! Schizophrenic molecules? Neurotic molecules? No way! The message is clear: the temptation to interpret phenomena that are inherently chemical in nature in biological terms—fitness, natural selection, adaptation, survival of the fittest, cooperation, information, etc., should be firmly resisted. Open any chemical text that deals with chemical reactivity and those biological expressions will not be found there. Chemical phenomena are explained in chemical (and physical) terms, as chemistry is the more fundamental science. On this basis a reinterpretation of Fig. 6 in terms of just one scientific discipline makes clear that the discipline of choice must be the lower-level one, chemistry, not the higher-level one, biology. So let us proceed to do just that. Let us reinterpret the entire process of Fig. 6—part chemical, part biological—solely in chemical terms.

Natural selection is kinetic selection

When several replicating molecules are mixed with their component molecular building blocks, as described in chapter 4, they compete with one another, in much the same way as biological entities compete for a limited supply of food. But as explained above we shouldn't discuss that competitive process as *natural selection at the molecular level*. Such reactions are dealt with by a specific branch of chemistry that deals with the rates of chemical reactions called chemical kinetics. That sub-discipline of chemistry, going back some 100 years to the pioneering work of Alfred Lotka, has no difficulty in dealing with the situation in which two replicating molecules compete for the same building blocks. It comes up with a clear-cut prediction that is applicable in most cases—the faster replicating molecule will out-replicate the slower replicating molecule and drive it to extinction. That result comes out directly by solving the relevant rate equations. In other words when two replicating molecules compete for the same chemical building blocks, the outcome is readily explained by a process that chemists call *kinetic selection*. Kinetic selection in everyday language just means 'the faster one wins'. Since the faster replicator is capable of assembling building blocks into new replicating molecules more effectively (for a variety of chemical reasons), the number of those faster replicators grows quickly while the number of slower replicators drops until those slower replicators disappear entirely.

But that strictly chemical result does ring a biological bell. It sounds very much like the way in which natural selection operates in biology. When two biological species compete for the same resource, the one that can utilize that resource more effectively drives the other to extinction. That result is the basis for the competitive exclusion principle that we discussed earlier. But then, natural selection and kinetic selection are really the same concept, so let us state that explicitly:

natural selection = kinetic selection

Biological natural selection merely emulates chemical kinetic selection. Natural selection is the biological term, kinetic selection is the chemical term.

At this point the reader may ask why the chemical description is to be preferred over the biological one. Despite the earlier comment that explanatory arrows always point downward, aren't the chemical and biological explanations really saying the same thing, that faster, and therefore more effective replicators, whether chemical or biological, will out-replicate less effective ones? Not quite. The reason is that the chemical explanation is more fundamental and probes the issue of selection more deeply. The chemical term is more quantifiable than the biological one because chemical systems are inherently simpler. That greater simplicity allows us to further break down the composite chemical replication reaction into the individual reaction steps that go to make it up. The chemical analysis can tell you how long it will take for one molecular replicator to out-replicate the other. It will even tell you under what circumstances the two replicators may coexist. Coexistence between competing molecular replicators can also be observed under appropriate circumstances.

Biological systems, on the other hand, are many orders of magnitude more complex, and are therefore less amenable to a detailed chemical analysis. That is why the two subjects are typically discussed at their different hierarchical levels. No matter, the recognition that natural selection has its roots within a fundamentally chemical phenomenon, one that is well understood, provides an important link connecting the two sciences of chemistry and biology.

Fitness and its chemical roots

What about that central biological term 'fitness'? What is the chemical analogue of that term and what new insights does the translation of that central biological term offer? According to Darwin, fitness is just the capacity to survive and reproduce, and its optimization is deemed the ultimate goal of the evolutionary process. Yet that concept, conceived by Darwin in strictly qualitative terms, has become a source of endless confusion due to continuing attempts to formally quantify it. The large number of fitness types that have been proposed and discussed—absolute fitness, relative fitness, inclusive fitness, ecological fitness, to mention some key ones—clearly attest to the inherent difficulties in this venture. The problem of fitness is a highly complex one, and one that has been troubling leading biologists for the better part of the past half-century, so a detailed discussion is well beyond the scope of this book. In the present context our goal is a more limited one: to explore how the merging of chemistry and biology can assist in clarifying at least some aspects of the troublesome 'fitness' issue.

In our earlier discussion of replicating systems we identified a fundamental characteristic of those systems—their dynamic kinetic stability, DKS. The ability of a replicating system to maintain itself over time reflects its stability, but a stability kind that differs from the conventional thermodynamic one. Our discussion now reveals that 'fitness' is actually the biological expression of that more general and fundamental chemical concept, so let us state that explicitly:

fitness = dynamic kinetic stability (DKS)

When we classify a biological entity as 'fit' we are really specifying that it is stable—stable in the sense of being persistent. However, as we explained previously in some detail, that stability kind only applies to a population, not to individual replicators within the population. Specifying that a population is fit (or stable) just means that the population is able to maintain itself through ongoing replication/reproduction. The immediate consequence of relating fitness and DKS is that it indicates more explicitly that fitness is best viewed as a population characteristic, not an individual one. The concept of DKS has no real meaning at the individual level. A stable population of some replicating system is the reality that comes about through individual replicators being formed and then decaying, like the water droplets turning over in a fountain. In the context of life, if you focus on the individual entity, tempting as it may be, you are missing the essence of what defines life—its dynamic nature, the continual turnover of the individual entities that make up a particular replicating population. Bottom line: in order to understand life's essence one should focus on life's *population* aspect, not its *individual* aspect. Life is an evolutionary phenomenon and evolution does not operate on individuals, only on populations. Individuals are just born and then die. Focus on the individual and you will miss much of what life entails. In actual fact the difficulty in individual thinking goes deeper than the above comments might suggest. What is an individual living entity, and do they actually exist? The answer to this question is more complex than we might imagine, but I will defer this aspect of the discussion to chapter 8.

The fact that a population perspective is crucial for a proper understanding of replicator dynamics received considerable impetus from important theoretical work carried out in the 1970s by Manfred Eigen, the eminent Nobel prize-winning German chemist, together with Peter Schuster, the distinguished Austrian chemist, on what is termed quasispecies theory.[58] In order to understand that theory in simplest terms we first need to describe what is meant by a 'fitness landscape'. As already discussed in chapter 4, when a replicating molecule, say an RNA of some particular sequence, proceeds to replicate, occasional errors in the replication reaction will result in the formation of RNA mutants. Mutants that are faster replicators will tend to drive the slower replicating sequences to extinction. That process of sequence modification can be represented by what is termed a fitness landscape—a three-dimensional topographical map. In that three-dimensional representation, the horizontal axes represent sequence changes (that come about through mutations) while the vertical axis represents the fitness of the particular sequence. The higher the value on that vertical axis, the greater the fitness. Accordingly, the fitness landscape resembles a three-dimensional topological map of mountain ranges and valleys in between. High points on the landscape—mountain peaks—represent RNA sequences of high fitness (fast replicators) and low points—valleys—represent RNA sequences of lower fitness (slower replicators). What that means is that some initial RNA sequence of a particular fitness, a point on that topology map, will tend to explore the fitness landscape in search of the highest point on the fitness landscape, representing the sequence of highest fitness, much like a hiker in the mountains seeking to climb to the top of the highest peak.

But here's the important point. Manfred Eigen and Peter Schuster discovered that the population of replicating RNAs that is generated by this exploration of the fitness landscape does not consist of one single sequence, but rather a *population* of RNAs of differing sequences, centred around the most successful sequence (termed the wild type) within that population. This population of varied sequences was termed a quasispecies, and an analogy that may help capture the essence of a quasispecies would be a flock of birds as it moves in concert over the sequence landscape in search of ever higher peaks. Eigen and Schuster discovered through their computer modelling of evolutionary changes in the RNA sequences that it is not the *fittest sequence* that is selected for but the *fittest population of sequences*—the fittest quasispecies—that is selected for. In other words, evolution operates by seeking out improved fitness in a *population* sense rather than in an individual sense. In fact one can see in Eigen and Schuster's seminal work the importance of population heterogeneity. A mutation leading to a particularly successful replicator is as likely to come from a *slower* RNA as from a *faster* one.

Counter-intuitively, the road to a fitter population may actually pass through a 'less fit' individual replicator within the existing population. Population heterogeneity opens up more possibilities for evolution to carry out its magic—heterogeneous populations evolve more effectively than homogeneous ones. The message is clear: the essence of stability in the world of replicators is rooted in populations, not individuals. Evolution is a process that populations undergo, not individuals. In the evolutionary scheme of things the individual is but a fleeting event, a transient water droplet in the fountain of life.

We have discussed the concept of DKS in some detail and the pertinent question now arises: can DKS be quantified? The short answer—only to a limited extent. We have identified DKS as a distinct stability kind in nature and stated that evolution operates so that DKS tends to increase over time. That fact alone suggests the term is quantifiable. Surely, if we say that evolution leads to greater DKS, that means that DKS is measurable. Yes and no. Take two RNA molecules in Sol Spiegelman's experiment competing for building blocks during self-replication and we see one replicates faster than the other. So the relative rates at which the two RNAs replicate may be taken as a quantitative measure of the relative DKSs of the two RNA populations. Recall, it is precisely because of this rate difference that a population of one replicator drives the other population to extinction. However that attempt at quantification is only of limited value for two reasons. First, attempts at quantification are only relevant for two populations that feed off a common resource. That means it can be applied to two RNA populations competing for the same set of activated nucleotides. But asking whether an *E. Coli* bacterium or a camel is more stable is meaningless, even in a population sense—there is no common frame of reference; it would be like comparing apples and oranges.

But the difficulty in quantifying DKS goes deeper. Despite the above comments implying that relative DKS *can* be estimated for competing replicators, another problem arises. The relative rates of replication depend on the reaction conditions. Let's return to Sol Spiegelman's classic RNA test-tube experiments. If some extraneous material is introduced into the test tube that inhibits the replication rates, as in fact was done by Spiegelman, then the winner of the replication race can switch. Change the reaction conditions and the evolutionary course changes as well; the winner of the Darwinian race is likely to be an entirely different set of RNA molecules. When Sol Spiegelman added a substance, ethidium bromide, to the reaction mixture, the winner of the Darwinian race turned out to be a different sequence.[59] Why is that? Since the extraneous material that had been added affected the mechanism of replication, certain sequences that initially facilitated rapid replication were inhibited, while other RNA sequences, initially slower, were favoured. In other words DKS for populations of RNAs is *circumstantial*. Its magnitude depends on the particular materials that are present in the reaction. But that means that DKS is quite different from that other kind of stability we speak about in chemistry, thermodynamic stability. The thermodynamic stability of water is a defined quantity regardless of what else is present (though to be precise, it does depend to some extent on the physical conditions, temperature, pressure, etc.). Thermodynamic stability is an intrinsic property of any system and is measured in *closed systems*. Dynamic kinetic stability depends on rates of reaction, is highly sensitive to reaction conditions, and can only be assessed in *open systems*, in which energy and resources are continually supplied. That makes comparisons of DKS highly problematic.

To clarify the point further, let us consider a biological example—a population of bacteria in a pool of water. Such a population may well be highly stable—billions of bacteria are busy replicating, resulting in the establishment of a dynamic population of those bacteria. But add chlorine to the pool and the bacteria simply die—their stability vanishes. The DKS of that bacterial population has dropped to zero. Same bacteria, different circumstances. The thermodynamic stability of the water molecules in that pool, however, does not depend on the environment. It is measured relative to that of a hydrogen-oxygen gas mixture (the materials from which water is formed), and that difference does not depend on the location of the water and the presence of other materials in the water (at least to any significant extent). Attempting to quantify DKS for a particular system is like preparing for an exam where the answers to the questions keep changing!

There is a moral to the above story: some characteristics of undeniable scientific interest are inherently difficult to quantify, or are even unquantifiable. Attempts to quantify the unquantifiable will be unrewarding and may only lead to confusion. The above discussion relating fitness to its underlying chemical term, DKS, makes clear why attempts to quantify fitness have proved so elusive. Not everything that counts can be counted. In addressing the concept of fitness, context is everything.

Despite the difficulties we've discussed in our ability to formally quantify DKS, two crude measures of DKS are actually available. These are the steady-state population number for a given replicating entity and the length of time that the replicating population has managed to maintain itself. A large steady-state population of some life form means that it is more readily able to withstand environmental changes that may undermine its existence. By contrast, if the population size of some living form is low, then clearly that population is vulnerable and may become extinct. On that basis it is reasonable to conclude that cockroaches and mosquitoes are more stable (in a DKS sense) than pandas. Cockroaches and mosquitoes are unlikely to become extinct in the foreseeable future, while the future of pandas is far less certain. Replication is ultimately a numbers game. The time dimension can also be a useful gauge of DKS. Cyanobacteria, which have maintained a continuing presence on our planet for several billion years, would of necessity be classified as stable, remarkably so. Modern humans, by comparison, have only existed for some 150,000–200,000 years, so our long-term stability is far less assured. No matter. Appreciating that fitness is the biological expression of a particular kind of stability helps place biology in a more physical context, and assists in our goal of merging the biological and physical sciences.

Survival of the fittest and its chemical roots

Having established the connection between fitness and stability we can now seek the chemical equivalent of that quintessential biological phrase 'survival of the fittest', or its more modern biological expression, 'maximizing fitness'. This will prove of importance since this connection will lead us to the profound realization that the entire evolutionary process illustrated in Fig. 6, i.e., both the emergence of life from inanimate beginnings, as well as the evolution of simple biological systems into more complex ones, may be associated with an identifiable driving force. The fact that a driving force for that entire process might be identifiable should not be a great surprise—that, after all, is nature's way. In nature many processes are associated with a driving force. Flowing rivers, rainfall, avalanches, falling apples, are all associated with the gravitational force, while the driving force for all chemical reactions is just the omnipresent Second Law of Thermodynamics. Of course the term 'force' needs to be interpreted broadly, in line with our comments on pattern recognition in chapter 3. Forces do not have to be visible to be identified. The existence of a force is postulated through empirical recognition of its action. In the case of replicating systems and their clear tendency to become transformed into more effective replicating systems, the driving force can now be identified as *the drive toward greater DKS*. In other words, the biological term 'maximizing fitness' is just the biological expression of the more fundamental and more 'physical' concept—maximizing DKS. So let's equate them:

maximizing fitness = maximizing DKS

In simple language that just means that replicating systems tend to become more successful replicators—those that can maintain themselves more effectively over time. In this light, abiogenesis and biological evolution, the two phases of Fig. 6, are *both* an expression of the drive toward greater DKS. Thus the transformation of inanimate matter to simple life should not be viewed as a collection of haphazard and contingent chemical events, but rather as a coherent process governed by an identifiable driving force. That same force operates during the so-called abiogenesis phase as well as the biological phase, and that is why the two phases should be viewed as one. The driving force for evolution is not natural selection, as is suggested from time to time. Natural selection is a rudder, not a driving force. Natural selection, as its name states, just selects. Natural selection (or its chemical equivalent, kinetic selection) helps steer the replicating population toward higher DKS by the continual elimination of those entities in the population that contribute to a lowering of its DKS. And that is true along the entire evolutionary road, from the population of simple (but unidentified) replicating entities which heralded life's tentative beginnings, right through to complex life.

We have specified the driving force for evolution, and already identified a primary mechanism for change—complexification. As we discussed earlier in this chapter, complexification is a main (but not sole) mechanism by which replicating systems enhance their DKS. That is not necessarily obvious if one looks at evolutionary changes on short time scales, just as one cannot see the ageing process in humans on a day-to-day basis. But if we step back and look at the evolutionary process over an extended period of time what we see is quite unmistakable. It is increasingly clear that life started off simple—some highly unstable replicating system (a replicating molecule or small replicating network)—and ended up complex, with multicell replicators—elephants and the like—roaming around.[60] That complexifying transformation was no fly-by-night operation—it took place over literally billions of years. But its undeniable presence all around us is unequivocal testimony to the reality of the process. From microscopic beginnings in some unknown location on the prebiotic earth, the process overwhelmed the entire planet with life forms of all sizes, all linked together in an awe-inspiring ecological network. Let us now consider that complexification process in more detail.

That initial replicating system on the prebiotic earth must have been highly fragile. It takes a sophisticated chemist in a sophisticated lab to induce a replicating molecule to replicate. Replicating molecules can be quite temperamental, and it is no easy matter to get them to replicate. Ask any systems chemist! But take any small sample of soil in your backyard, look carefully, and what you find are billions of bacteria replicating away quite happily—no chemist in charge, no lab, no postdocs, no equipment. Bacteria are highly robust and effective replicators. Bacteria are highly stable in a DKS sense and that high stability has come about through the long evolutionary process of complexification. In other words, a step-by-step process of complexification led from some replicating system, initially fragile and highly unstable, to the highly stable, highly complex, replicating entity that is a bacterial cell. We don't know and will likely never know the identity of that first replicating system or its first tentative steps toward robust life—that historic information is shrouded in the mists of time. However the *nature* of the first step on that long road to complex biological systems is clearly illustrated in Gerald Joyce's two-molecule RNA system which replicated more effectively than any single RNA molecule.[57]

Information and its chemical roots

The concept of information permeates all of modern biology. In fact entire books have been written on the topic, and for good reason. The concept of information, as manifest in the sequence of the nucleotide building blocks that make up the DNA molecule, has been central in our understanding of the key processes of molecular biology. In that sense DNA replication can be thought of as the process that provides for the preservation of information, one in which genetic information is passed on from generation to generation. And how is this information expressed? Transcription and translation, those key processes on which the central dogma of biology rests, deal precisely with that issue—the manner in which the information in the DNA sequence is somehow translated into the multitude of unique protein structures on which much of life's functionality is based. But here's the puzzle. Open a chemical textbook and you will most likely not find the 'information' word in there at all. Chemists talk about reactivity, selectivity, stability, reaction rates, catalysis, and many other chemically useful terms, not about information. So how can that be? If biology is just a complex kind of chemistry, and information is central to biology, then information must exist within chemistry as well. If not, where did the information within biological systems come from?

To answer the above questions, let us now think about DNA's reactions in chemical rather than biological terms. That wondrous DNA molecule, through the process of replication, acts as an *auto-catalyst*. DNA is an autocatalyst because it catalyses its own formation. The building blocks that go to make up a second copy of DNA will not proceed to do so unless one DNA molecule is already present, one that serves as a template, and thereby facilitates that autocatalytic behaviour. But, of course, DNA does not just catalyse its own formation. Through the processes of transcription into messenger RNA and the subsequent translation of the messenger RNA sequence into an amino acid sequence (proteins), it also acts as a *catalyst*, a catalyst for the synthesis of other materials. In the absence of that DNA molecule with its specific sequence, the transformation simply could not occur. The precise sequence of the DNA segment that is expressed through operation of the ribosomal machinery determines the precise structure of the protein that is obtained. Change the DNA sequence and you end up with a different protein structure. What that means therefore is that DNA is not just an autocatalyst, *but also a highly specific catalyst*. A particular DNA segment sequence corresponds to a particular protein, a different sequence would correspond to a different protein. A moment's thought suggests therefore *that the term 'information' in its biological context is just 'specific catalysis' when considered in a chemical context*. The different jargons employed by the two sciences can create divisions that are not actually there. So in a chemical context what has taken place over the evolutionary time frame is that some autocatalyst endowed with catalytic properties evolved in such a way as to enhance those catalytic properties. Initially that primordial nucleic acid autocatalyst might have just exhibited the ability to catalyse the formation of simple peptides as a precursor to some primitive translation process. However, over time, and with the establishment of the genetic code, the efficiency and the specificity of that catalytic capability would have continually increased, driven by the DKS imperative. The biological phenomenon of information generation is nothing other than the chemical phenomenon of *establishing and enhancing specific catalytic function*.

A comment regarding the connection between 'information' and the Second Law is now called for. The highly ordered sequence of the DNA molecule is of course thermodynamically unstable. The Second Law dictates that ordered systems tend to become disordered so information tends to be degraded, not created. That might suggest that the generation of information out of nowhere would contradict the Second Law. But of course there is no contradiction. Just as my writing of this book creates information (hopefully), the process of evolution can also create information, provided, of course, the appropriate energy cost is paid. That's where life's metabolic processes come in—to supply the required energy to keep life's machinery going and maintain life's far-from-equilibrium state. So the question of how information emerged from non-information is just a rephrasing of Schrödinger's question of how unstable, far-from-equilibrium systems, emerged in the first place. We will deal with that question shortly.

Toward a general theory of evolution

Based on the previous arguments we can now piece together the central elements of a general theory of evolution, one that is applicable to both chemical replicating systems as well as biological ones. Like its biological counterpart, its central elements revolve around replication, mutation, and selection, but as we have indicated, the process of complexification needs to be incorporated into the general scheme so that the causal sequence in evolution becomes: *replication, mutation, complexification, selection, evolution.*

Let us then begin by specifying the *why* and the *how* of evolution. The *why* is the driving force for the process—the drive toward greater replicator DKS. The *how* is the mechanism for that process and would have comprised the steps previously mentioned—replication, mutation, complexification, selection. The process is initiated by the emergence of some oligomeric replicating entity susceptible to imperfect replication. An RNA molecule, or one related to it, illustrates the kind of molecule that would have been able to initiate the process, though other possibilities cannot be excluded. In any case, its precise identification would not be necessary in seeking the principle of the process. Once that molecule begins to self-replicate, either on its own or within a minimal network, it would tend to enhance its stability (of the dynamic kinetic type, as we have described earlier) due to the driving force that operates within the world of replicating entities—the drive toward greater DKS. And now to the second step. Replication occurs with occasional mutations thereby creating a diversity of replicators. Moreover, if one includes horizontal gene transfer as an additional mechanism leading to genetic diversity, then it is apparent that genetic variation does not have to derive solely from the replication step.

And now to the complexification step. Any molecular replicator (or replicating network) once formed would tend to interact with other available materials potentially leading to more complex replicators. Importantly, that process of complexification would have been initiated from the outset, at the molecular level, the moment the system was governed by that other stability kind, DKS. Of course, complexification would not need to be restricted solely to members of the class of replicating molecules. Replicating sequences capable of catalysing the formation of *other* chemical classes, e.g., peptides, exhibiting catalytic activity with respect to the replication reaction itself, would further add to the process of complexification and evolution toward more stable replicating systems. Complexification would therefore entail a *co-evolutionary process* in which non-replicative molecules could also partake in the building up of increasingly complex replicative networks. Such a process would continue unabated as long as the system as a whole remained holistically autocatalytic. Thus it is complexification, through the establishment of increasingly complex chemical networks, that would be the primary mechanism for the enhancement of replicator DKS and the generation of stable replicating systems.

And finally selection. Once a population of diverse replicating systems is established then (kinetic) selection would act to change the proportion of replicators within the population to those able to better contribute to the population's DKS. Of course the result of that process of continuing cycles of replication, mutation, complexification, selection, is *evolution*.

Let us now return to the issue of what drives evolution. Our earlier discussion, where 'fitness maximization' has been translated into 'DKS maximization' helps place biology squarely in the physical-chemical world where ultimately it should belong. Just as in the 'regular' chemical world the drive of all physical and chemical systems is toward the most stable state, in the replicative world the drive is also toward the most stable state, but of the kind of stability applicable within that replicative world, DKS. We see then that the material world can in some sense be subdivided into two parallel worlds—the 'regular' chemical world and the replicative world. Transformations in the 'regular' world are governed by the Second Law, and in the replicative world by what could be considered to be an analogue of the Second Law. So, as is manifestly evident, we live simultaneously in two discrete chemical worlds—two worlds governed by different kinds of stability and therefore expressing two quite distinct chemistries. As we have seen, one of these chemistries, the chemistry of the replicative world, is called biology.

BIOLOGY IS CHEMISTRY

How did a metabolic (energy-gathering) capability come about?

But now the unavoidable question must be asked. How can there be two laws that govern chemical transformation? Isn't this a contradiction? How can there be two kinds of stability? What happens when these two kinds of stability pull in opposite directions? Which would win? The answer is quite surprising. Even though the Second Law is the ultimate law, the one whose directive cannot be ignored, it is actually the Second Law analogue that wins! Let's see how this comes about. In doing so we will obtain insight into the issue that troubled Erwin Schrödinger—how could far-from-equilibrium systems have emerged naturally?

It is true that no physical or chemical system can undergo change contrary to the strict requirements of the Second Law. To do so would be equivalent to proposing that balls spontaneously roll uphill, and they don't. However, if a replicating system were to acquire an energy-gathering system, then nature could have its cake and eat it. It would be the existence of such a system that would enable the drive toward greater DKS to comfortably coexist with the strict requirements of the Second Law, despite the often opposing requirements of these two stability kinds. But how could this come about naturally?

In a recent theoretical simulation, Emmanuel Tannebaum and Nathaniel Wagner, two colleagues in the chemistry department at Ben Gurion University, and myself, have demonstrated that a replicating molecule that underwent some chance mutation that enabled it to capture energy, say, light energy, in a primitive kind of photosynthesis, would be able to out-compete a molecular replicator that lacked such a capability and drive it to extinction.[61] This could even be true if the energy-gathering replicator was intrinsically slower! How can that be? The process of replication requires that the building blocks from which the molecular copy is composed be chemically activated. Activation is necessary to enable the building blocks to link up once they have locked into place on the template molecule. That's a Second Law requirement and it must be obeyed. However activated (high-energy) building blocks are likely to be in short supply compared to unactivated (low-energy) ones. So a non-metabolic replicator (without an energy-gathering capability) would quickly use up the available quantity of activated building blocks at which point the replication reaction would cease. If, however, the replicating molecule is metabolic (i.e., it possesses an energy-gathering system), then that replicating molecule, by acquiring energy, could transmit that energy to the building blocks that have attached to it, thereby activating them. In other words, the existence of an energy-gathering capability within the replicator molecule can effectively *increase* the availability of activated building blocks, thereby facilitating the replication reaction for the metabolic replicator.

The more general point is that the existence of an energy-gathering capability within a replicating entity effectively 'frees' that entity from the constraints of the Second Law in much the same way that a car engine 'frees' a car from gravitational constraints. A motorized vehicle is not restricted to merely rolling downhill, but thanks to an external energy source (gasoline), can travel uphill as well. In other words, just as a motorized vehicle is a more effective vehicle for travel, so a replicator that can gather energy will likely be a more successful replicator than one that cannot. The importance of the simulation described above is that it demonstrates that a replicating system that acquires *an energy-gathering capability by a chance mutation would be more stable in a DKS sense and would therefore be selected for over one without that capability.* Until now we had considered structural complexification as the primary way of enhancing DKS, but we can now see that complexification of a different kind—metabolic complexification (in the energy-gathering sense)—could also have the same DKS enhancing effect. In fact, *the moment some non-metabolic (downhill) replicator acquired an energy-gathering capability, could be thought of as the moment that life began.* At that moment the replicating system would be free to pursue its replicating 'agenda' despite associated energy costs, and, significantly, through the incorporation of that energy-gathering system the conflicting requirements of DKS and the Second Law would be accommodated. The means by which *thermodynamically unstable*, but *DKS stable* entities could emerge is clarified. The problem that troubled Erwin Schrödinger and other physicists would seem to have a feasible solution.

Metabolism or replication first?

In the light of our discussions on the origin of life and the theory of life presented, we can now reassess the 'metabolism first–replication first' dichotomy, a question that has plagued the origin of life debate for several decades now. As we will now see, the unification of abiogenesis and biological evolution as one process may largely resolve the uncertainty at the base of that debate. The underlying issue is whether template replication or a primitive metabolism (simple autocatalytic cycle formation) would have been the central element in the emergence of life. Recall, the 'replication first' school considers a chain-like molecule, such as RNA, capable of replication through a template-type mechanism, as the source of life's emergence on earth, while the 'metabolism first' school believes that a simple (holistically) autocatalytic cycle would have been necessary in order to create a self-replicating system. A consideration of Gerald Joyce's insightful experiments on RNA replication provides a hint as to the feasible resolution of the issue. Recall, a single RNA molecule was unable to replicate in a robust way, but a two-molecule network *was* able to do so. This key result suggests that both template-directed autocatalysis *and* cycle formation may well have been critical elements in the emergence of life, *most likely closely synchronized*. Simply put, *complexification (i.e., the establishment of reaction cycles) could not have come about without replication, and template replication without complexification had nowhere to go*. We are suggesting then that the 'replication first–metabolism first' dichotomy, as a fundamental issue in the origin of life debate, may no longer be of real relevance, and that the two conflicting approaches should be replaced by a bridging *replication and metabolism together* scenario. This Solomon-like resolution suggests that replication *and* the emergence of a primitive metabolism (a simple autocatalytic cycle) were both crucial elements at the very earliest stages of life's emergence. Only in combination was life able to emerge from its simple inanimate beginnings.

8

What is Life?

We have presented many pieces of a highly intricate puzzle—the life puzzle, and in this final chapter we will attempt to piece the puzzle together and outline a theory of life that can offer answers to Schrödinger's simple 'what is life' question. The test of the theory will be relatively easy. It would need to explain in simple chemical terms why life has the special properties and characteristics that it has, to clarify the principles that would explain the process by which it emerged from non-life, and at least attempt to offer a broad strategy for its synthesis from its molecular building blocks.

Before summarizing the elements that make up our theory of life, we must not forget that there exists a well-established theory of matter—quantum theory, a theory that in principle at least can predict the properties and future behaviour of *any* chemical system. That might suggest that a theory of life should just be part of that more general theory. Formally it is, but in a way that makes the life issue inaccessible. In practice quantum theory can only deal with chemical systems of moderate size, and biological systems in their totality are way too complicated to be treated in that fundamental way. Go to a computational quantum chemist and ask him to solve a biological problem that requires him to explicitly treat the system's full complexity and he will likely just grimace and walk away. So a separate theory of life is needed. Separate, but not independent. The theory of life described here is but a smaller Russian doll within the bigger 'theory of matter' doll. Importantly, however, and as emphasized earlier in the text, *despite* life's complexity, a theory of life can be postulated, with the basis for such a theory being the presumption that life began simple, and that life's essence reflects its simple beginnings. By probing what we believe to be the equivalent of life's simple beginnings, we are able to grasp biology's core and address some of biology's most basic questions. But to do that, to get to the core, we had to cut through the many layers of complexity to uncover what lies hidden inside, and that was done by peeling away the layers of complexity *along a reverse time axis*. Complexity was built up over time, step by step, so we had to conceptually reverse that process until the core was reached. Only by reaching back into the process by which life on earth emerged can the essence of what it is to be alive be uncovered. Once we get to that core, we may begin to understand *why* life emerged, and have a clearer view of *what* life is.

That approach leads us to systems chemistry, the chemistry of simple replicating systems that we discussed in detail in earlier chapters. The study of simple replicating systems has revealed an extraordinary connection—that Darwinian theory, that quintessential biological principle, can be incorporated into a more general chemical theory of evolution, one that encompasses both living and non-living systems. It is that integration that forms the basis of the theory of life I propose. The realization that chemistry and biology connect up in this fundamental way will prove, we believe, to have profound implications, some of which are already apparent, for example, the unification of abiogenesis and biological evolution. Abiogenesis and biological evolution are one continuous process—abiogenesis (the transformation of non-living matter to earliest life) is the low-complexity phase, biological evolution is just the high-complexity phase. That unification serves to clarify the physical nature of the evolutionary process that led from simple abiotic beginnings right through to complex life. By uncovering the process that connects inanimate to animate, the essence of what it is to be alive begins to materialize. The emergence of life was initiated by the emergence of a simple replicating system, because that seemingly inconsequential event opened the door to a distinctly different kind of chemistry—replicative chemistry. Entering the world of replicative chemistry reveals the existence of that other kind of stability in nature, the dynamic kinetic stability of things that are good at making more of themselves. Exploring the world of replicative chemistry helps explain *why* a simple primordial replicating system would have been expected to complexify over time. The reason: to increase its stability—its dynamic kinetic stability (DKS).

Yes, living systems involve chemical reactions, lots of them, but the essence of life, the process that started it all off, was replication. And what makes that replication reaction special is not *what* it produces but *how much* it produces. If a further reminder of the special nature of the replication reaction is needed, consider a single replicating molecule, weighing just 10^{-21} grams. If it were to replicate once a minute, then, in under five hours that replicating molecule would have grown (in principle, of course) into a mass exceeding that of the entire universe! Think about that! One molecule replicating not too rapidly, would devour the entire material resources of the universe in a few hours! The point is that the replication reaction is unique and totally different from every other chemical reaction that appears in a chemistry textbook because of that awe-inspiring kinetic power—a mathematical power that turns the conventional rules of chemistry on their heads. The Second Law of Thermodynamics is, of course, fully applicable to replicating systems, but the enormous kinetic power of replication ends up seemingly circumventing that ubiquitous Second Law. The concept of stability in chemistry is fundamental, but that extraordinary kinetic power creates a distinctly different kind of stability in chemistry from the ones we are familiar with. As discussed in chapter 4, in 'regular' chemistry matter is stable if it *doesn't* react. But in the world of replicating systems, matter is stable (in the sense of being persistent) if it *does* react, to make more of itself. And in this persistent sense, matter that is *better* at making more of itself is more stable than matter that isn't.

That is the essence of the DKS concept. But that means that in the world of replicators, reactions follow a *Second Law analogue*—populations of poorer replicators continually react so as to become more effective (more stable) replicators, though, of course, only in a manner that is consistent with the Second Law itself. And the kind of chemistry that results from reactions in this 'other world', the replicative world, is so different from those in the 'regular' world that much of it goes under a different name—biology. Biology then is just a particularly complex kind of replicative chemistry and the living state can be thought of as a new state of matter, the *replicative state of matter*, whose properties derive from the special kind of stability that characterizes replicating entities—DKS. That leads to the following working definition of life: *a self-sustaining kinetically stable dynamic reaction network derived from the replication reaction.* Each word in the definition imparts an important element to the definition. 'Self-sustaining' means that the system must have an energy-gathering capability in order to satisfy the requirements of the overriding Second Law. The terms 'kinetically stable' and 'dynamic' describe the characteristics of that other stability kind, and the words 'network' and 'replication' are self-explanatory, though we will shortly expand on the network aspect of life, one of considerable importance. Of course, from that perspective, death is just the reversion of a system from the kinetic, replicative world back to the thermodynamic world, the world of 'regular' chemistry.

So there we have it. Even though life is an extraordinarily complex phenomenon, the life principle is surprisingly simple. Life is just the resultant network of chemical reactions that emerges from the continuing cycle of replication, mutation, complexification, and selection, when it operates on particular chain-like molecules—in the case of life on Earth, the nucleic acids. It is possible that other chemical systems could also exhibit this property, but so far this question has yet to be explored experimentally. Life then is just the chemical consequences that derive from the power of exponential growth operating on certain replicating chemical systems.

The theoretical ideas at the heart of the DKS concept are far from new. Thomas Malthus fully appreciated the mathematical power of exponentials, as described in his classic work 'An Essay on the Principle of Population' published in 1798, and Alfred Lotka's early work on kinetic theory going back to 1910 fully appreciated the kinetic consequences of exponential growth on both chemical and biological systems. Paradoxically that is all the 'hard theory' one needs to know to understand life. Note, no quantum mechanics involved—that murky area of physics and chemistry that continues to strain human credulity. In that sense life is a *classical* phenomenon and the tendency of past physicists to attribute life's character to matter's fundamental quantum nature appears unnecessary. Though the importance of quantum effects in many areas of chemistry is undisputed, it is surprising how much organic chemistry and biochemistry is understandable without the need for quantum thinking. It is the *complexity* of life that has created confusion and blocked important early insights, particularly those of Malthus, Lotka, and Troland, and more recently, those of Manfred Eigen and Peter Schuster. So the relationship between the life phenomenon and its extraordinary complexity can now be stated: complexity is not the *cause* nor the *essence* of the life phenomenon, complexity is its *consequence*. Replication induced complexity, not the other way around. It is the coupling of long-standing and basic theoretical ideas associated with autocatalytic systems together with the insights from recent studies of simple replicating systems, and the networks they establish, that enables the elements of the life puzzle to be finally pieced together.

Of course any theory is only as useful as the range of phenomena it can explain. In the following pages we will revisit the life characteristics that we discussed in chapter 1—its complexity, its teleonomic character, dynamic character, its diversity, its far-from-equilibrium state, and its chiral character, to see how the theory of life we have offered can explain these properties. Finally, as the scientific method requires, I will make some predictions that flow directly from the theory of life that has been outlined.

Understanding life's characteristics

Life's complexity

Life's extraordinary, almost incomprehensible complexity was described in chapter 1 and we can now see that understanding the nature of DKS explains that extraordinary complexity. And as we have already discussed, the mechanism by which nature enhances DKS is through complexification—not complexification in the sense of aggregation, which we routinely see in the 'regular' chemical world, but one that is quite different, and is unique to the replicative world. When materials aggregate in the 'regular' chemical world—for example water freezing into ice or any solid crystallizing out of solution—that process happens because the solid aggregate is the more stable form. But that stability kind is thermodynamic stability, the stability kind associated with being *less* reactive, the kind that we are so familiar with in chemistry. All the physical aggregates that we generally see in the world around us derive from that simple idea—the molecules that make up those aggregates attract one another resulting in aggregates that are more stable, and hence less reactive, than the separated molecules.

But in the replicative world the stability kind that is applicable is DKS, so the aggregation pattern that is observed is the one that enhances *that* stability kind, not thermodynamic stability. And while that aggregation process will almost certainly have thermodynamic contributions to it, those contributions are secondary, and merely facilitate the primary one, which directs toward enhanced DKS. We met that interaction at its very simplest level when we discussed Gerald Joyce's striking RNA experiment in which two

RNA molecules catalysed each other's formation, thereby leading to the establishment of a small replicating network. In simplest terms, once a simple and relatively fragile (meaning unstable in DKS terms) entity comes about, it will tend to complexify in order to enhance its DKS. It is that Woody Allen 'whatever works' rule in operation again. The process occurs step by step, each step leading to a slightly more complex entity capable of enhanced replicative ability. As we noted earlier, that early process would have most likely consisted of an expanding chemical network of reactions whose overall character would be replicative—a replicating network. One can only speculate as to the specific steps that took place along the long road to early life, but the drive toward greater DKS through the mechanism of increasing complexity would characterize the process. So the above analysis couched in DKS terms explains why stability in the replicative and 'regular' chemical worlds are distinct, and why the aggregation processes in each of the two worlds, in particular during the process of life's emergence, necessarily follow different paths. After several billion years of evolution the end product can be understood—replicators whose complexity is one of staggering proportions, even in simplest life, and also of extraordinary stability (in DKS terms). High complexity and high DKS go hand in hand.

As a final point, and as already noted earlier, in some instances a process of simplification, rather than one of complexification, is observed during evolution, and at both chemical and biological levels. It is that 'whatever works' idea again—in biology there are few hard and fast rules. Nature has no objection to taking an evolutionary step of simplification, if such a step enhances a replicator's DKS. Whatever works! It is the DKS maximization principle that enables evolutionary processes at both chemical and biological levels to be understood.

Life's instability

We have already noted that all living things are unstable in a thermodynamic sense, like a bird constantly flapping its wings to maintain its airborne state. And just like that hovering bird, all living things must constantly consume energy to maintain that far-from-equilibrium state. Yet, somehow the world is totally overwhelmed with these thermodynamically unstable entities. How come? Shouldn't unstable things gradually disappear, rather than continue to be formed and take root in just about every feasible ecological niche? But, based on our discussion in chapter 4, all living things actually *are* stable, but their stability is of that 'other kind'—DKS, the stability of things that are good at making more of themselves. As already stated, in the world of replicators the stability that matters is DKS and not thermodynamic stability. And why is it that those entities that are stable in a DKS sense are invariably *unstable* in a thermodynamic sense? Simply, because DKS depends on the system continually reacting in order to replicate, to make more of itself, and that actually requires the system to be reactive, to be unstable. Thermodynamically stable entities don't react. They are like balls at the bottom of a slope—they have nowhere lower to roll. In other words for a living system to be a highly successful replicator it has to be DKS *stable* and thermodynamically *unstable*. We discussed how these two seemingly contradictory requirements can be simultaneously accommodated when a replicating system acquired an energy-gathering capability through a process of kinetic selection. Replicators that have an energy-gathering capability are better replicators than those that don't—just like cars *with* an engine are more useful forms of transport than cars *without*. Once a replicator with an energy-gathering capability came about by some chance mutation, being of higher DKS (a more effective replicator) it quickly drove its predecessor into extinction. That's why all living systems, with no exception, have an integrated energy-gathering system in place—the photosynthetic one in the case of plants and certain bacteria, and the Krebs (citric acid) cycle for the catabolic breakdown of organic matter in the case of animals. The result: the world is full of DKS *stable*, but thermodynamically *unstable*, replicating systems. These two stability kinds, potentially in opposition to one another, can live together harmoniously thanks to that energy-gathering capability. Recently Robert Pascal, an innovative French chemist, has begun to explore the kinds of chemical processes that would have facilitated the emergence of early metabolic systems, during the transition to modern metabolic pathways.[62]

Life's dynamic nature

One of life's striking characteristics is its dynamic nature. We commented earlier how within the space of some months you are no longer who you were. Materially you are now largely composed of new stuff—a new you! Your blood cells, billions of them, are replaced daily, your skin cells continually turn over, the protein molecules that do most of the work in getting on with life are all continually being degraded and regenerated in a never-ending dynamic process. But how can this ephemeral and dynamic nature of living systems be explained? In fact, this particular aspect of life is one of the easiest to understand. Recall our analogy of a replicating population and a water fountain. The fountain is stable (persistent) even though the water that makes up that fountain is turning over continuously. Different water, same fountain. For any replicating entity the same proposition holds. Because the replication reaction is unsustainable, regardless of what it is that is being replicated, a replicating system that achieves stability would be one in which the rate of replicator generation and decay would be in rough balance, one in which a steady state is established. This would be true of molecules, microbes, or monkeys, or any other replicating entity one would care to mention. In other words it is the population that is stable, with the individual entities that make up that population constantly turning over. And this continual turnover holds at all levels of complexity—molecules within cells are constantly turning over, cells within organisms are constantly turning over, and, of course, all organisms are constantly turning over. That simple fact clarifies the role that death plays in the life process. In a 2005 commencement speech to Stanford graduates, Steve Jobs, the hi-tech innovator said:

> No one wants to die. Even people who want to go to heaven don't want to die to get there. And yet death is the destination we all share. No one has ever escaped it. And that is as it should be, because Death is very likely the single best invention of Life. It is Life's change agent. It clears out the old to make way for the new. Right now the new is you, but someday not too long from now, you will gradually become the old and be cleared away. Sorry to be so dramatic, but it is quite true.

Death then is not just something bad that happens to us living things. Death is part of the life strategy. Seeking eternal life? The term is an oxymoron. There can be no eternal life because the very basis of life is its transient and dynamic nature.

WHAT IS LIFE?

Life's diversity

Though Darwinian theory was able to relate all living things to one another, the source of life's spectacular diversity remains unresolved. As we discussed in chapter 1, Darwin himself remained uncertain on this key point. In his *Origin of Species* Darwin did propose a Principle of Divergence, but whether that principle was independent of his principle of natural selection, or derived from the principle, was left open and, interestingly, the issue continues to preoccupy modern biologists. However, the theory of life that we have described, based on the DKS concept, seems to offer some resolution of this issue. It turns out that the key to understanding life's extraordinary diversity lies in the *topologies* of the two chemical worlds—the 'regular' and replicative worlds, and the difference between them. Let me explain.

I have already explained that all chemical systems are directed toward their most stable form. That means that different chemical systems that are composed of the same elements will all want to end up at the same place, just like different balls rolling down a hilly

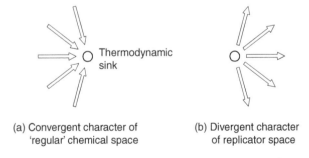

(a) Convergent character of 'regular' chemical space

(b) Divergent character of replicator space

Fig. 7. Schematic representation of branching patterns within 'regular' chemical space (convergent), and within replicator space (divergent).

terrain from different locations on that terrain will all want to end up in the same location—the lowest point in the valley below. If you take any mixture of hydrocarbons—that's just a chain of carbon atoms joined to hydrogen atoms, such as we find in gasoline—and react that mixture with oxygen in what is called a combustion reaction, the resultant product is carbon dioxide and water. It doesn't matter which hydrocarbon you start with, you always end up with carbon dioxide and water, because that is the most stable form of a mixture of C, H, and O atoms. All hydrocarbon-oxygen mixtures converge to carbon dioxide and water. That argument may be generalized to chemical systems as a whole, so one could say that the grid that connects the world of 'regular' chemical substances is *convergent*, as illustrated schematically in Fig. 7a. All roads lead to Rome and all chemical reactions are directed to what is called their thermodynamic sink—the lowest energy possibility for that combination of atoms. That's how a chemist can frequently predict the result of a chemical reaction, that's how he/she knows where the chemical system 'wants to go'.

But let us now turn to the world of replicating systems. In contrast to a 'regular' chemical system, which may be thought of as contained, or closed, a replicating system must remain open at all times to allow the replicating reaction to proceed unimpeded. Being open means that building blocks for replication, as well as the energy to support the replication process, must be continually provided. In other words, in comparison to a 'regular' chemical reaction, which may be carried out in a closed container, a replicating reaction must remain open to the surroundings. That different situation results in the path to greater DKS being *divergent*, as illustrated in Fig. 7b, rather than convergent. Why? Because the path forward to greater DKS will depend on what's available at that time and in that place, and any number of different paths toward more stable systems (in a DKS sense) are, in principle, feasible. Some replicator X might pair up with some molecule Y to create a more stable X/Y system compared to X on its own, but it also might pair up with some other molecule Z, thereby creating a stable X/Z system. The possibility of different complexification pathways leads to diversification. All stable replicating systems are continually replicating, occasionally mutating, continually complexifying, thereby exploring the world of replicating systems for increasingly effective replicators. The topology of the world of replicating systems is inherently divergent.

This different topology for the two worlds has interesting consequences. It not only explains life's diversity but it also explains how we are able to go back in time and seek our evolutionary roots. A divergent topology in the forward direction becomes a *convergent* one in the backward direction. It is that convergent topology in the *reverse* direction that enables us to utilize phylogenetic analysis and the fossil record to trace our evolutionary history going back in time, to deduce that all living things can be divided into three life kingdoms—Archaea, Bacteria, and Eukarya—to trace out the history of life on earth toward life's so-called Last Universal Common Ancestor (LUCA). But that, of course, means that we can say nothing at all regarding where evolution may take us in the future. Set off on a divergent path and there's no telling where you'll get to. As Yogi Berra, the well-known sports celebrity, once put it: 'If you don't know where you are going you will wind up somewhere else.' The different reactivity patterns of both 'regular' and replicative systems as a function of time—forward or backward—is simply explained.[63,64]

Life's homochirality

We have remarked how life's homochiral (single-handed) nature presents a puzzle at two levels. First, how did life's single-handedness emerge from a universe that is inherently two-handed, and second, how is that homochirality maintained, given that homochirality is intrinsically less stable than heterochirality. We have seen in this book that one of the key ideas that can explain the emergence of life on earth is the enormous kinetic power of autocatalysis. It is then remarkable to discover that the unexpected emergence of homochirality from a heterochiral environment can be explained in precisely the same terms! Normally when one carries out a chemical reaction that transforms a non-chiral substance (possessing no handedness) into a chiral one, the product is composed of equal amounts of left- and right-handed forms. But in 1995 the renowned Japanese chemist, Kenso Soai, made a remarkable discovery.[65] In certain instances it is possible to obtain effectively just one homochiral product from a non-chiral starting material. Somehow the symmetry of the system is broken, which is quite extraordinary. It's like tossing a coin a thousand times and observing 999 heads and one tail! No wonder Kenso Soai's unexpected result caused a sensation. In other words it *is* possible to generate homochiral systems, starting from a non-chiral environment, even though for many years this was considered physically unreasonable. So what has this to do with the emergence of life?

Soai's highly unexpected result is explained by the fact that the chemical reaction he studied proceeds *autocatalytically*, and therefore product formation shows exponential growth. If the reaction mixture is initially seeded with a tiny excess of one of the chiral products, then the spectacular amplification that autocatalysis can generate results in that product reaching a level of purity very close to 100 per cent. In other words, just as replication is autocatalytic, so homochirality (single-handedness) can be induced in a system that shows autocatalytic behaviour. This reaction and its detailed explanation are somewhat technical but the bottom line is straightforward: *the kinetic power of replication which is responsible for the emergence of life could well have been responsible for one of life's most striking features—its homochiral character.* The pieces of the life puzzle do fit together. How satisfying!

We have explained the *emergence* of homochirality from a nonchiral environment, but how is that homochirality maintained if homochirality is intrinsically less stable than heterochirality. Like several previous life dilemmas, this issue is also resolved through the DKS concept. Yes, systems that are of one chiral form *are* less stable than heterochiral mixtures, but that is only true in a thermodynamic sense. We have already seen that in the context of replicating systems, the stability that counts is DKS, and for this stability kind it turns out that homochiral systems are *more* stable than heterochiral ones. Life's reactions require high specificity, meaning precise lock-and-key type interactions between reacting molecules and that can only be obtained in homochiral systems. Introduce heterochirality into such systems and you will end up with half the keys not fitting into their locks! Homochiral systems are therefore more effective replicators than heterochiral ones, and as a consequence homochiral systems exhibit greater stability in the crucial DKS sense.

Life's teleonomic character

We discussed this most amazing of life's properties in some detail in chapter 1. To reiterate, both the structure and the behaviour of all living things lead to an unambiguous and unavoidable conclusion—living things have an 'agenda'. Living things act on their own behalf. But how can that be? How can matter, when organized in the manner we classify as biological, seemingly follow different rules from those of inanimate systems? How can matter of any kind appear to have an agenda? Let us see how the DKS concept can help resolve this puzzle. Recall that the reactions of simple replicating systems—say, replicating molecules—would follow the thermodynamic directive, much like a car without an engine follows the gravitational directive—it can only roll downhill. But once a replicating entity has taken on an energy-gathering capability, the replicating entity is now 'freed' of thermodynamic constraints and can follow the *kinetic* directive—the drive toward greater DKS. As we discussed earlier, a replicating entity with an energy-gathering capability is now like a car *with* an engine—it can go uphill too. That means that a replicating system with an energy-gathering capability would *appear* to have an agenda. It would seem to be acting purposefully, as it would no longer need to be confined to the downhill thermodynamic path, which we interpret as *objective* behaviour, but rather the path toward systems of greater DKS, which could involve the equivalent of rolling some way uphill. In other words, once a replicator has taken on an energy-gathering capability (as part of the general process of complexification toward more complex and more stable replicating systems), we would interpret and understand its subsequent replicative behaviour as *purposeful*.[66] Monod's paradox—how a purposeful system can emerge from an objective universe, is seen to result from the interplay of kinetic and thermodynamic directives in chemical reactions. In the 'regular' chemical world, thermodynamics is the dominant directive and results in so-called *objective* behaviour. In the replicating world, kinetics is the dominant directive and so actions in that world *appear* purposeful.

Consciousness

There are other profound life issues that we have not touched upon—consciousness, for example. While consciousness is certainly a characteristic of life, it is not an essential one, as it is only associated with advanced life forms. Accordingly, we have not dealt with it. Nonetheless, the issue of consciousness should be mentioned, if only to demonstrate how limited our understanding of some life characteristics remains. Having said that, the phenomenon of consciousness can be explored through its evolutionary context. Evolution is the process by which all properties of matter are exploited in the evolutionary drive toward more effective replicating systems. Evolution exploits matter's propensity for hardness when that is useful, as in bones. It exploits matter's ability to be flexibly firm when that is needed, as in cartilage; matter's ability to be liquid when that is needed, as in blood; matter's ability to be transparent as in crystallin, the protein from which the lens of the eye is made; matter's ability to conduct electric charge, and so on. But it turns out that matter in some particular organization has an even more remarkable characteristic—the remarkable property of consciousness. Indeed, an extraordinary characteristic—matter can be self-aware. Evolution has discovered that capability of matter, like all others that it has come across, and utilized it in the ongoing search for stable replicating entities. If we want to understand consciousness and its basis, we should study its source—neural activity at its most rudimentary level, and then track the phenomenon, step by step, through to its more advanced manifestations, ultimately to us humans. So the approach would be the same as the one we have taken in addressing the problem of abiogenesis—start simple. A fascinating scientific journey awaits us.

How would alien life look?

Having explained life's global characteristics in chemical terms, we can now pose the question: how would alien life look? Since we believe that life on Earth emerged from inanimate matter, it naturally follows that under appropriate conditions life could also emerge elsewhere in the universe. And while that life could also be based on the same molecular foundation—the nucleic acid–protein duo—other replicative combinations cannot be ruled out. We now understand that the basis of life consists of long-chain molecules capable of catalysing their own replication, which together with other chain-like molecules possessing catalytic capabilities would undergo a continual process of replication, mutation, and complexification. However, there is no reason at all to believe that in principle there would not be chemical combinations, other than that nucleic acid-protein duo, that could lead to that same general result. In fact, all of our experience in chemistry tells us that chemical characteristics are related to groups of substances, not to individual ones, so the expectation would be that, in principle at least, there would be a *group of materials* on which the processes of life could be based. So, if life did emerge on some other planet, one based on a different biochemistry from that on earth, can our theory of life offer some insight into how such life would appear? I believe so. My short answer: life on other planets would look exactly like that on our own!

I write that partly tongue in cheek because life's diversity has offered us an unimaginably large array of forms, from microscopic bacteria through to blue whales, so it is hard to see how life forms of any other kind would strike us as fundamentally different in their external appearance, and certainly no more alien looking than many of life's existing forms. More to the point, however, is the fact that life's morphology appears to be based on what living things require it to be, rather than some directive that comes from its underlying chemistry. Cars made from fibreglass, aluminium, or steel don't look too different from one another because their appearance is based on the shape cars need to be in order to function as cars. All cars, regardless of the material from which they are made, require an external shell in which to house the motor and create a cabin for passengers to sit in. They all possess windows so the driver can see where he is going, and wheels to minimize friction. That is true whether the cars are made in the US or in China, whether the windows are glass or plastic, whether the engine is electric or gasoline. In the same way, life forms that emerged from some replicating entity that did not belong to the nucleic acid family, but were able to complexify and evolve toward replicating entities of greater DKS, would likely utilize the same universal concepts that nucleic acid-based biochemistry discovered. Depending on the extent to which that other life form had evolved, it would also express network characteristics, and may have discovered the replicative value of a cell structure, in which the cell's functional parts with its replicative and metabolic capabilities would be incorporated. The theory of life presented here is not one based on material, but one based on process, and therefore the nature of the material would be secondary, possibly even incidental, in governing life's underlying characteristics.

Synthesizing life

Which brings us to the most intriguing of questions—how would one synthesize a simple living system? To this question there is no simple answer. If the theory of life presented here teaches us anything, it is that the synthesis of some entity that would possess the characteristics of a primitive life form, say a protocell, faces enormous difficulties. Let's see what these are. I will begin with some observations.

The relationship between living and non-living systems is particularly fascinating in at least one respect. It is so easy to transform living systems into non-living ones, but, as we know all too well, the process is not reversible—life is so easy to destroy, but (chemically speaking) so hard to make. That simple fact in itself is highly informative. The problem with the synthesis of a living system is not one of material, but, as noted, one of organization. You can have all the components of a living cell available, but packaging it so that it behaves as a living entity is where the difficulty lies. So what is the problem? Life is a *dynamic* state of matter meaning that the biomolecules that make up the living cell are in a constant state of flux. A simple physical analogy that captures this dynamic character would be that of a juggler juggling several balls. That dynamic state is of course identical in a material sense to the one in which a man stands next to those balls, which are resting on the ground. But the difference is profound. How easy it is to take a juggler juggling several balls and to convert him into the non-juggling state, one in which all the balls are lying on the ground. A hefty push and you are there! A man standing next to five balls would be the metaphor for death. Of course going in the other direction is not that simple. You cannot simply throw five balls in one go at a person and expect him to enter the juggling state. That won't work. In the same way, if you take all the components of a living cell and mix them together, you won't end up with a living cell. At very best, if all the bits and pieces end up in the right place, you'll end up with the equivalent of a *dead* cell. You'll end up with a clump of stuff—a thermodynamic aggregate. Recall, however, that the living cell is in a dynamic, far-from-equilibrium state, like that bird flapping its wings to stay airborne. Simply bringing together the components that can potentially make up an integrated and dynamic system that we would classify as alive won't lead to that special organizational and dynamic character that we recognize as life.

So let us return to our juggler analogy to see what kind of strategy might work. How does one enter the juggling state? The answer—step by step. Initially you toss two balls at the juggler, then a third, then a fourth, one step at a time. You start off simple and you add complexity bit by bit. That's how evolution did it—step by step, from simple and less stable, to complex and more stable. So how to make life? Enter the replicative dynamic state at a low level of complexity and then proceed to complexify, one step at a time. That, of course, is easier said than done. But don't be fooled by morphology. Life, even in its very simplest form, is far more than just a replicating entity in a bag. If a much simpler individual life form were capable of a physical existence, then it stands to reason that we would see such life forms as part of the replicative array of possibilities, as part of the passing parade, but we don't. The absence of such entities speaks volumes for their physical feasibility. Given life's dynamic nature, the synthesis of a chemical system expressing the dynamic characteristics of life would be an important step forward. Recently Sijbren Otto, an innovative systems chemist from the University of Groningen with a Ph.D. student, Elio Mattia, have begun exploring possible means of generating such dynamic kinetically stable chemical systems, but the challenges are great. I will conclude by saying that the synthesis of a simple chemical aggregate that exhibits lifelike characteristics, primarily self-sustained replication, appears to be a highly ambitious target at the present time.

How did life emerge?

We stressed early in this book that if we want to understand what life is, we have to understand the process by which it emerged. And what have we discovered? That thanks to recent findings in systems chemistry, the origin of life problem, at least in its ahistoric sense, may be largely resolved. There is now good reason to believe that abiogenesis and Darwinian evolution are just one process. So, if we believe we understand biological evolution, and broadly speaking we do, then we also understand abiogenesis. The historical questions—the *what*, *where*, and *when* questions, will continue to tease and torment us for the foreseeable future, as the ability of scientific study and reasoning to uncover the historical record is limited. However, just as the historical details of Darwinian evolution—

what species lived when—are secondary to the theoretical framework, so the historical details of life's emergence, though fascinating in their own right, could also be considered of secondary importance. A solution to the primary question exists and is breathtakingly simple: life on earth emerged through the enormous kinetic power of the replication reaction acting on unidentified, but simple replicating systems, apparently composed of chain-like oligomeric substances, RNA or RNA-like, capable of mutation and complexification. That process of complexification took place because it resulted in the enhancement of their stability—not their thermodynamic stability, but rather the relevant stability in the world of replicating systems, their DKS. What is particularly satisfying in this explanation is that the resolution of the origin of life problem (in the ahistorical sense) dovetails seamlessly with Charles Darwin's momentous ideas on biological evolution. In effect the physical problem of how life on earth emerged may be understood by reformulating and extending Charles Darwin's theory of biological evolution to include molecular systems. By reinterpreting and translating the central biological terms that underlie biological evolution into the corresponding chemical terms, it becomes evident that abiogenesis and biological evolution are indeed one single chemical process.[64]

Of course, as we pointed out above, that explanation does not tell us what actual events took place on the earth 4 billion years ago. But then Darwin's theory of evolution does not tell us the specific historic path from earliest life to today's diverse and complex life either. That wasn't its purpose. Filling in the historic record was left to palaeontology and phylogenetic analysis. Darwin's contribution was in delineating ahistoric principles. He revealed to us that biological evolution is a natural process, that all living things are related and descended from some common ancestor, and that a simple mechanism, natural selection, operating on mutating replicators, can explain the basis for that entire process. What has been argued here is that the Darwinian thesis can be extended to inanimate matter enabling the problem of abiogenesis to be resolved in the same ahistorical manner. It is staggering to realize that Darwin, in his genius, already foresaw where his evolutionary principles might ultimately lead. His comment in his 1882 letter to George Wallich 'that the principle of continuity renders it probable that the principle of life will hereafter be shown to be part, or consequence, of some general law' seems almost clairvoyant in its precision and clarity. Darwin didn't know about replicating molecules or kinetic selection or the mechanism for biological hereditary or those insightful experiments in systems chemistry of Gerald Joyce, Günter von Kiedrowski, Reza Ghadiri, Gonen Ashkenasy, Sijbren Otto, and other fine chemists, but well over a century after those words were written it seems Darwin was, yet again, right on the mark.

Life as a network

Having clarified the central elements in the process of life's emergence from inanimate matter, we are now ready to address a fascinating and central feature of living things, one that dramatically impacts on life's very essence—its network character. We have already seen that life began simple and then proceeded to complexify. But what do we actually mean by 'complexify'? The answer: network formation—from relatively simple reaction networks through to complex ones. The essence of all these networks is that they are holistically self-replicating. Life then is just a highly intricate network of chemical reactions that has maintained its autocatalytic capability, and, as already noted, that complex network emerged one step at a time starting from simpler networks. And the driving force? As discussed in earlier chapters, it is the drive toward greater DKS, itself based on the kinetic power of replication, which allows replicating chemical systems to develop into ever-increasing complex and stable forms. And now the actual nature of that complexification process can be specified—network formation. Complexification, network formation—they are effectively one and the same. Viewed in this light, life is more a *process* than it is a *thing*. Or as Carl Woese and Nigel Goldenfeld recently put it: 'Biology is a study, not in being, but in becoming.'[67] And in what medium does that network establish itself? In that extraordinary solvent, water. Water, the cosmic juice, with its unique properties[68,69] is considered crucial for enabling life's network of reactions to have become established.

I have stated that life is a network of chemical reactions, but merely by inspecting the world around us we see that the network seems to be composed of individual units—cells. Cells are the smallest discrete entities that we unambiguously term to be 'living'. Living things can consist of these single-cell entities, or they can be multicell organisms composed of blocks of individual cells. But the network perspective on life leads to an interesting and highly pertinent question: *Do individual life forms actually exist?* Individual living things do seem to exist, in the sense that we are surrounded by what appear to be examples of individual life forms—birds, bees, camel, humans, and, of course, unicell life, primarily bacteria, all seemingly going about their individual business. In practice, however, that individuality is not quite as clear-cut as one might think. What we classify as individual living entities may themselves be thought of as components of a network—the ever-expanding life network. Let's think again about those single-cell species, bacteria. We discussed earlier that in some bacterial species the colony can actually switch from a unicell format—swarms of individual bacteria—to a multicell format, where the bacteria merge together into a protoplasmic lump. That happens when resources become scarce. But those cells, whether bound together or physically separate, are constantly communicating chemically to coordinate their actions. The phenomenon of biofilms is another example of coordinated bacterial action. Bacterial behaviour highlights life's network character. Bacterial genes destine them to be *communal*, not *individual*. Bacteria are more a network than they are a set of individuals.

Recent thinking regarding the evolutionary path that led to the modern cell fits into this general mould. Carl Woese considers early cells to have been highly communal, their evolution dominated by horizontal gene transfer.[70] That is another way of saying that earlier life consisted of a tightly integrated replicative network of simpler aggregates. But, as the network evolved and complexified further, it advanced to a looser and more modular form. That's when cells, as discrete biological entities, were born. That transition was a highly significant one—one might consider it as a phase transition.[55] That morphological change from strictly communal to increasingly individual opened up a new range of evolutionary capabilities. One obvious advantage of that transition was that a replicating network whose components exhibit greater individual character would be less vulnerable to attack than a tighter interdependent network.

Attack any segment of a tight network, in which all components are crucial for network replication, and the entire network will suffer. If, however, the network is made up of components that are themselves replicative, then the network can be looser and more modular. Destroy some components of a looser, more modular, network and the network is likely to survive. *But that means that individuality is more a life strategy than a life characteristic.* So-called individuality is just a technique that evolution has discovered, amongst many others, to enhance replicative ability and robustness. This network perspective can change the very way we think about life, and reaffirms that the life phenomenon is better understood as one of *process* rather than one of *form*, the forms being incidental manifestations of the process. Looked at in this way the life process—the replicative process—can be seen to utilize every 'trick in the book' in order to optimize its replicating agenda. The process chooses togetherness when that is optimal, and separateness, manifested as physical individuality, when that is the better option. Whatever works best at the given time and under the particular circumstances.

What about the role of individuality in multicell systems? Surely here one could argue that clear-cut and unambiguous cases can be recognized. However, here also that individual classification is quite problematic. Take us humans as an example. Each human is, of course, composed of billions of individual cells, some 10^{13} of them, and of many different kinds. Remarkably however, each human being actually consists of ten times as many bacterial cells as human ones. From a numerical point of view, we are more bacterial than human! Literally billions of these bacteria, comprising hundreds of different species, reside in our gut, in other body cavities, on our skin. Each human is more a superorganism—a giant network—

than an organism.[71] These bacteria may be so integral to human health that they have recently been described as the 'forgotten organ'![72] The point is that each and every human individual, and so every multicell creature, is more an ecological network than a single living entity. Indeed, appreciating life's inherent network character, rather than focusing on its individual character, is leading us to a new and revolutionary way of understanding disease and disease prevention, at least when viewed from the human perspective.

What about plants? Plant individuality is also questionable as they also are part of an extensive ecological network. Plants depend on bacteria for their metabolism much like animals, though by a different mechanism. Plants depend on a source of nitrogen to enable protein synthesis, but atmospheric nitrogen is relatively inert and cannot be utilized readily. It is the bacteria in the soil and in the plant's roots that enable plants to access nitrogen in a usable form. We see then that life is more like a set of Russian dolls nestled in one another, and connected up in networks with other sets of dolls, rather than an extensive array of independent things that interact with one another. Even those bacteria that inhabit your gut are not the last link in the replicative chain, but may themselves be hosts to lesser life forms, viruses. Viruses are non-metabolic entities that are only able to replicate by exploiting the metabolic capabilities of their host cells. Are viruses then the end of the line? In life there are always unexpected surprises. It has recently been discovered that giant viruses are abundant in nature, some larger in size than small bacteria. Interestingly, however, it has recently been found that these large viruses can themselves be infected—with small viruses. As with Russian dolls, you are never sure when you have reached the last link in the chain. Replicative chemistry is full of unexpected twists and turns.

That was the network looking down, but start from a human and look up and you see an individual who is part of a nuclear family, which, in turn, is part of an extended family, which is part of a local community, which is part of larger groups of human organization. The functioning of the network at any level is dependent on the functioning of the network both below and above. The individual merely represents a particular level of complexity within a network that involves many different levels of complexity. Take sex, for example. It catches our attention—it's meant to. Sex tells us that we, as sexual individuals, are reproductively speaking incomplete. Biologically speaking, our individuality is actually non-existent. The individual has no future—literally. That's why sex catches our attention in that powerful and compulsive way. But we are also emotionally incomplete and various psychological elements also connect us to the network. We obsessively need to be with others. We think we are separate, but we are one. We think of ourselves as individuals, but we are really just components of a network. So a biosphere that has overwhelmed our planet should not be interpreted in terms of an invasion by billions of individual life forms, but by an ever-expanding living network. The replicative drive leaves no stone unturned in seeking novel and creative ways to replicate and extend that network. Clearly, given the above comments, coming up with a precise definition for an individual living thing would be problematic. Would an individual have to be reproductively independent? If so, any sexual being, like you or me, would not satisfy the definition. Would a life form be considered truly individual, if it is symbiotically bound to other replicating entities, without which it cannot reproduce or even survive? Even though some components of that giant replicative network do appear to be individual, that appearance is often illusory.

The network perspective on life might assist in addressing some of the questions concerning life that have been frequently raised over the years. Based on the theory of life proposed here, replication is the essence of life. That might seem to imply that a mule or lone rabbit would not be considered alive, as neither can reproduce. But, of course, mules (and lone rabbits) are alive. It is true that they cannot reproduce but they *are* still part of the replicative network— they are just dead-ends. A road that stops in a dead-end is still a road and part of the road network. Mules *are* replicative entities, not because they can reproduce—they can't—but because of the replicative process by which they came into being. What about viruses—are they alive? One can conduct lengthy debates on the matter and ultimately the answer would depend on one's precise definition of a living thing. Clearly viruses are lacking key life characteristics, such as possessing an independent metabolism. Having said that, however, there is no doubting that viruses are also an integral part of the life network. For viruses the question is more philosophic and linguistic than scientific.

The merging of chemistry and biology

The goal of this book has been to demonstrate that answers to several of the most central of life questions, including the classic one posed by Schrödinger, are finally becoming accessible. The extraordinary powers of science and the inductive method in particular, have revolutionized our lives and our understanding of the world to an extent we could not have foreseen, even a century ago. Thanks to the remarkable scientific progress these past 150 years, from Darwin's awesome revolution in biological thinking, through to the exciting new developments in systems chemistry, biology and chemistry are finally merging, finally becoming one. The Darwinian revolution may now be nearing its ultimate goal, the one that Charles Darwin already foresaw 130 years ago—the integration of the biological sciences within the physical sciences. That merging of the two sciences means that within the limits that science itself imposes on us, we can begin to understand what is life, why it emerged, how we, a twiglet on the tree of life, together with all other living things on our planet, relate to the material world and the universe as a whole, and why, despite the unforgiving harshness of the Darwinian view, we are committed to one another, why in some deeper sense, we are one. Can that fundamental life connection serve as a ray of hope for the future of humankind, the entity that Stephen Hawking called 'a chemical scum on a moderate-sized planet'? Only time will tell.

REFERENCES AND NOTES

1. Woese CR, A new biology for a new century. *Microbiol. Mol. Biol. Rev.* 68: 173–86, 2004.
2. Dawkins R, *The Blind Watchmaker*. Norton: New York, 1996.
3. Gold T, The deep, hot biosphere. *PNAS* 89: 6045–9, 1992.
4. Proctor LM, Karl DM, A sea of microbes. *Oceanography* 20: 14–15, 2007.
5. Soshichi U, Darwin's principle of divergence. <http:/philsci-archive.pitt.edu/1781/1/PrDiv.pdf>, 2004.
6. McShea DW, Brandon RN, *Biology's First Law: The Tendency for Diversity and Complexity to Increase in Evolutionary Systems*. University of Chicago Press: Chicago, 2010.
7. Haeckel E, *Die Radiolarien (Rhizopoda Radiaria): Eine Monographie*. Druck und Verlag Von Georg Reimer: Berlin, 1862; cited in Pereto J, Bada JF, Lazcano A, Charles Darwin and the origin of life. *Orig. Life Evol. Biosphere* 39: 395–406, 2009.
8. Bohr N, *Nature* 131: 458, 1933; cited in Yockey HP, *Information Theory, Evolution, and the Origin of Life*. Cambridge University Press: Cambridge, 2005.
9. Schrödinger E, *What is Life?* Cambridge University Press: Cambridge, 1944.
10. Monod J, *Chance and Necessity*. Random: New York, 1971.
11. Watson JD, Crick FH, Genetical implications of the structure of deoxyribonucleic acid. *Nature* 171: 964–7, 1953.
12. Popper K, Reduction and the incompleteness of science. In Ayala F, Dobzhansky T, eds., *Studies in the Philosophy of Biology*. University of California Press: Berkeley and Los Angeles, 1974.
13. Crick FHC, *Life Itself*. Simon and Schuster: New York, 1981.

14. Popa R, *Between Necessity and Probability: Searching for the Definition and Origin of Life.* Springer: Berlin, 2004.
15. Dyson FJ, Colloquium at NASA's Goddard Space Flight Center, 2000.
16. Kunin V, A system of two polymerases—a model for the origin of life. *Orig. Life Evol. Biosphere* 30: 459–66, 2000.
17. Arrhenius G, Short definitions of life. In Palyi G, Zucchi C, Caglioti L, eds., *Fundamentals of Life*, 17–18. Elsevier: New York, 2002.
18. Hennet RJC, Life is simply a particular state of organized instability. In Palyi G, Zucchi C, Caglioti L, eds., *Fundamentals of Life*, 109–10. Elsevier: Paris, 2002.
19. Cleland CE, Chyba CF, Defining life. *Orig. Life Evol. Biosphere* 32: 387–93, 2002.
20. Macaulay TB, *Critical and Historical Essays*, vol. iii. Project Gutenberg eBook #28046, 2009.
21. Ayala FJ, Dobzhansky T, eds., *Studies in the Philosophy of Biology: Reduction and Related Problems.* Macmillan: London, 1974.
22. Noble D, *The Music of Life.* Oxford University Press: Oxford, 2006.
23. Brenner S, Sequences and consequences. *Phil. Trans. R. Soc. B* 365: 207–12, 2010.
24. Weinberg S, *Dreams of a Final Theory.* Vintage: New York, 1994.
25. Crick FHC, *Of Molecules and Men.* University of Washington Press: Seattle, 1966.
26. Cornish-Bowden A, *Perspectives in Biology and Medicine* 49: 475–89, 2006.
27. Mills DR, Peterson RL, Spiegelman S, An extracellular Darwinian experiment with a self-duplicating nucleic acid molecule. *PNAS* 58: 217, 1967.
28. von Kiedrowski G, A self-replicating hexadeoxynucleotide. *Angew. Chem. Int. Ed. Eng.* 25: 932–4, 1986.
29. von Kiedrowski G, Otto S, Herdewijn P, Welcome home, systems chemists! *J. Syst. Chem.* 1: 1, 2011.
30. Dawkins R, *The Selfish Gene.* Oxford University Press: Oxford, 1989.
31. Grand S, *Creation.* Harvard University Press: Cambridge, MA, 2001.

32. For recent comprehensive reviews on the origin of life see: (a) Luisi PL, *The Emergence of Life: From Chemical Origins to Synthetic Biology*. Cambridge University Press: Cambridge, 2006; (b) ref. 14; (c) Fry I, *The Emergence of Life on Earth*. Rutgers University Press: Piscataway, 2000.
33. Wacey D, Kilburn MR, Saunders M, Cliff J, Brasier MD, Microfossils of sulphur-metabolizing cells in 3.4-billion-year-old rocks of Western Australia. *Nature Geoscience* 4: 698–702. doi:10.1038/ngeo1238, 2011.
34. Mojzsis SJ, Arrhenius G, McKeegan KD, Harrison TM, Nutman AP, Friend CRL, Evidence for life on Earth before 3800 million years ago. *Nature* 384: 55–9, 1996.
35. Hanage WP, Fraser C, Spratt BG, Fuzzy species among recombinogenic bacteria. *BMC Biology* 3: 6, 2005; Papke RT, Gogarten JP, How bacterial lineages emerge. *Science* 336: 45, 2012.
36. Woese C, Interpreting the universal phylogenetic tree. *PNAS* 97: 8392–6, 2000.
37. Miller SL, A production of amino acids under possible primitive earth conditions. *Science* 117: 528–9, 1953.
38. Waechtershaeuser G, Groundwork for an evolutionary biochemistry: the iron-sulphur world. *Prog. Biophys. Mol. Biol.* 58: 85–201, 1992.
39. Cairns-Smith A, *Genetic Takeover and the Mineral Origin of Life*. Cambridge University Press: London, 1982.
40. Powner MW, Gerland B, Sutherland JD, Synthesis of activated pyrimidine ribonucleotides in prebiotically plausible conditions. *Nature* 459: 239–42, 2009.
41. Szostak JW, Bartel DP, Luisi PL, Synthesizing life. *Nature* 409: 387–90, 2001.
42. Kauffman SA, *Investigations*. Oxford University Press: Oxford, 2000.
43. Dyson FJ, *Origins of Life*. Cambridge University Press: London, 1985.
44. Eigen M, *Steps toward Life: A Perspective on Evolution*. Oxford University Press: Oxford, 1992.
45. Gesteland RF, Atkins, JF, *The RNA World: The Nature of Modern RNA Suggests a Prebiotic World*. Cold Spring Harbor Laboratory Press: Cold Spring Harbor, NY, 1993.

46. Lifson S, On the crucial stages in the origin of animate matter. *J. Mol. Evol.* 44: 1–8, 1997.
47. Orgel LE, The implausibility of metabolic cycles on the prebiotic earth. *PLoS Biol.* 6: e18, 2008.
48. de Duve C, *Life Evolving: Molecules, Mind and Meaning.* Oxford University Press: Oxford, 2002.
49. Ganti T, Organization of chemical reactions into dividing and metabolizing units: the chemotons. *BioSystems* 7: 189–95, 1975.
50. Prigogine I, Time, structure and fluctuations. *Science* 201: 777–85, 1978.
51. Collier J, The dynamics of biological order. In Weber BH, Depew DJ, Smith JD, eds., *Entropy, Information, and Evolution*, 227–42. MIT Press: Cambridge, MA, 1988.
52. Gardner M, Mathematical games: the fantastic combinations of John Conway's new solitaire game 'Life'. *Scientific American* 223: 120–3, 1970.
53. Voytek SB, Joyce GF, Niche partitioning in the coevolution of two distinct RNA. *PNAS* 106: 7780–5, 2009.
54. Hardin G, The competitive exclusion principle. *Science* 131: 1292–7, 1960.
55. Maynard Smith J, Szathmary E, *The Major Transitions in Evolution.* Oxford University Press: Oxford, 1995.
56. Dadon Z, Wagner N, Ashkenasy G, The road to non-enzymatic molecular networks. *Angew. Chem. Int. Ed.* 47: 6128–36, 2008.
57. Lincoln TA, Joyce GF, Self-sustained replication of an RNA enzyme. *Science* 323: 1229–32, 2009.
58. Eigen M, Schuster P, *The Hypercycle: A Principle of Natural Self-Organization.* Springer-Verlag: Berlin, 1979.
59. Saffhill R, Schneider-Bernloehr H, Orgel LE, Spiegelman S, In vitro selection of bacteriophage Qβ ribonucleic acid variants resistant to ethidium bromide. *J. Mol. Biol.* 51: 531–9, 1970.
60. Of course, not *all* life complexified over the evolutionary time frame. Microbial life was, and has remained, the most ubiquitous life form. The point is that from a world that was initially populated solely by relatively simple life forms, the evolutionary process did lead to the emergence of highly complex forms.

61. Wagner N, Pross A, Tannenbaum E, Selection advantage of metabolic over non-metabolic replicators: a kinetic analysis. *BioSys.* 99: 126–9, 2010.
62. Pascal R, Boiteau L, Energy flows, metabolism and translation. *Phil. Trans. R. Soc. B* 366: 2949–58, 2011; Pascal R, Suitable energetic conditions for dynamic chemical complexity and the living state. *J. Syst. Chem.* 3: 3, 2012.
63. Pross A, Stability in chemistry and biology: life as a kinetic state of matter. *Pure Appl. Chem.* 77: 1905–21, 2005.
64. Pross A, Toward a general theory of evolution: extending Darwinian theory to inanimate matter. *J. Syst. Chem.* 2: 1, 2011.
65. Soai K, Shibata T, Morioka H, Choji K, Asymmetric autocatalysis and amplification of enantiomeric excess of a chiral molecule. *Nature* 378: 767–8, 1995.
66. Pross A, How can a chemical system act purposefully? Bridging between life and non-life. *J. Phys. Org. Chem.* 21: 724–30, 2008.
67. Woese CR, Goldenfeld N, How the microbial world saved evolution from the Scylla of molecular biology and the Charybdis of the modern synthesis. *Microbiol. Mol. Biol. Rev.* 73: 14–21, 2009.
68. Engberts JBFN, in Lindstrom UM, ed., *Organic Reactions in Water: Principles, Strategies and Applications.* Wiley-Blackwell: London, 2007.
69. Lynden-Bell RM, Conway Morris S, Barrow JD, Finney JL, Harper Jr. CL, eds., *Water and Life: The Unique Properties of H_2O.* CRC Press: Boca Raton, FL, 2010.
70. Woese CR, On the evolution of cells. *PNAS* 99: 8742–7, 2002.
71. Gill RG, Pop M, DeBoy RT, Eckburg PB, Turnbaugh PJ, Samuel BS, Gordon JI, Relman DA, Fraser-Liggett CM, Nelson KE, Metagenomic analysis of the human distal gut microbiome. *Science* 312: 1355–9, 2006.
72. O'Hara AM, Shanahan F, The gut flora as a forgotten organ. *EMBO reports* 7: 688–93, 2006.

INDEX

abiogenesis 126, 182
Allen, Woody 50, 167
alien life 178
Altman, Sydney 105
archaea 3, 89
Aristotle 32, 33
autocatalysis 62–5, 68, 151

bacteria 90
bacterial diversity 23, 24
Bohr, Niels 36
Brenner, Sydney 53

catalysis 61, 62, 151, 152
Cech, Thomas 105
chemical reactions 58
chemotaxis 15–16
chirality 27, 28
Chyba, Christopher 41
Ciechanover, Aaron 22
Cleland, Carol 41
competitive exclusion principle 128
complexity 4
consciousness 177
Conway, John 119
Cornish-Bowden, Athel 57
Crick, Francis 54, 55, 83
cyanobacteria 74, 75

Darwinian theory 8, 34, 35, 112, 113, 117, 183, 184
Dawkins, Richard 4, 76
death 170
De Duve, Christian 108
definition of life 40, 164
Delbrück, Max 88
dissipative structure 118
diversity 171
DNA 38, 69, 151
dynamic kinetic stability (DKS) 73, 75, 78, 141, 144–6, 149, 150, 164, 166–9, 172

dynamic stability 71
Dyson, Freeman 103

earth's age 87
Eigen, Manfred 142, 143
Einstein, Albert 47
entropy 62
eukarya 90

Feynman, Richard 47, 101
finches, Darwin's 129
fitness 140, 141, 147, 148
fitness landscape 142

game of life 119
Ganti, Tibor 115
general theory of evolution 153
Grand, Steve 76

Haeckel, Ernst 35
Haldane, J.B.S. 83
Hawking, Stephen vii, 191
Hershko, Avram 22
hierarchical reduction 53, 137
holism 50–7
homeostasis 6
homochirality 28, 29, 174, 175
horizontal gene transfer 91
human genome project 113

induction 43
information 150–3

Jobs, Steve 170
Joyce, Gerald 128, 129, 132–4, 159, 166

Kauffman, Stuart 102, 114
kinetic selection 138, 139
kingdoms of life 90

INDEX

Last Universal Common Ancestor (LUCA) 88, 91, 173
Lifson, Shneior 107
Lotka, Alfred 138, 164
Luisi, Pier Luigi 99

Macaulay, Thomas 45
Malthus, Thomas 164
maximizing fitness 148
metabolism first 102, 106, 159
Miller, Stanley 93
molecular evolution 128
molecular replication 65–70, 77
Monod, Jacques 33, 36, 37, 51, 52, 108, 117, 176
mutation 78

natural selection 138, 149
Newton, Isaac 44–7
Noble, Denis 52
non-equilibrium thermodynamics 118
nucleotides 97

Oparin, Alexander 83, 93
Orgel, Leslie 96, 107
Otto, Sijbren 182

palaeobiologic record 88
panspermia 83
Pascal, Robert 169
pattern recognition 44
phylogenetic analysis 88
Popa, Radu 39
Popper, Karl 38, 52
prebiotic chemistry 92, 95
Prigogine, Ilya 117
Principle of Divergence 24, 171
prokarya 90
protein degradation 22

quasispecies 142

RNA 65–8, 78, 79, 102, 128–130, 132–5, 142, 143, 151
RNA-world 96, 104, 105

Ramakrishnan, Venkatraman 2
reduction 50–57
replication 72, 76, 79
replication first 102, 104, 105, 158, 159
ribosome 1
Rose, Erwin 22
Russian doll metaphor 80

Schrödinger, Erwin vii, 36
Schuster, Peter 98, 142, 143
scientific method 43
Second Law of Thermodynamics 6, 25, 59, 61, 79–81, 107, 152, 155, 156, 157, 163
self-replication 65–8
simplification 135, 167
Soai, Kenso 174
Spiegelman, Sol 65, 78, 127, 128, 135, 145
stability 71–5
Steitz, Thomas 2
Sutherland, John 97
systems biology 116
systems chemistry xii, 75, 76, 101, 123
Szostak, Jack 99

Tannenbaum, Emmanuel 156
thermodynamic stability 145
teleology 9, 33, 34
teleonomy 9–20, 37, 176
theory of life 42
tree of life 89–92

viruses 188
von Kiedrowski, Günter xii, 69

Wagner, Nathaniel 156
Weinberg, Steven 48, 53, 54, 137
Whitesides, George 82
Wittgenstein, Ludwig 48
Woese, Carl viii, 3, 51, 52, 89, 91, 112, 114, 185, 186

Yonath, Ada 2